Annals of Mathematics Studies

Number 62

# GENERALIZED FEYNMAN AMPLITUDES

BY

Eugene R. Speer

PRINCETON UNIVERSITY PRESS
AND THE
UNIVERSITY OF TOKYO PRESS

PRINCETON, NEW JERSEY

1969

L.C. Card: 72-77595

Published in Japan exclusively by the
University of Tokyo Press;
in other parts of the world by
Princeton University Press

Printed in the United States of America

*Acknowledgements*

I would like to thank my advisor, Prof. Arthur Wightman, for giving generously of his time during the past two years. His help and advice were invaluable in the preparation of this thesis. I would also like to thank Prof. Edward Nelson for reading the manuscript and for various suggestions, and Prof. Tullio Regge for several helpful discussions.

I am grateful to the National Science Foundation for support during four years of graduate school, to the Princeton University Mathematics Department for support during the summer of 1968, and to Dr. Carl Kaysen for his hospitality at the Institute for Advanced Study.

Part of this work was sponsored by the Air Force Office of Scientific Research, Office of Aerospace Research, United States Air Force, under AFOSR Grant 68-1365.

*ABSTRACT*

Renormalization in the context of Lagrangian quantum field theory is reviewed, with emphasis on two points: (a) the Bogoliubov-Parasiuk definition of the renormalized amplitude of an arbitrary Feynman graph, including some generalizations of the rigorous work of Hepp, and (b) a discussion of the implementation of this renormalization by counter terms in an arbitrary interaction Lagrangian. A new quantity called a generalized Feynman amplitude is then defined. It depends analytically on complex parameters $\lambda_1, \ldots, \lambda_L$, and these analytic properties may be used to define renormalized Feynman amplitudes in a new way; the method is shown to be equivalent to that of Bogoliubov, Parasiuk, and Hepp. The generalized Feynman amplitude depends on other parameters also; when these take on certain values, it is equal to the Feynman amplitudes for various graphs (aside from problems of renormalization, which are handled via the $\lambda$ dependence). The generalized amplitude thus interpolates the Feynman amplitude between different graphs. Some partial results are obtained which exploit this interpolation to give an integral representation for a sum of Feynman amplitudes.

## TABLE OF CONTENTS

## *INTRODUCTION*

The basic subject matter of this thesis is the *Feynman amplitude* (sometimes called a Feynman integral) which is always associated with a *Feynman graph.* A Feynman graph G is a graph [see Definition A.1] together with an assignment of a *propagator* $\Delta^{(\ell)}$ to each line $\ell$ of the graph; $\Delta^{(\ell)}$ is a distribution in $\mathcal{S}'(R^4)$ whose Fourier transform has the form

$$\tilde{\Delta}^{(\ell)}(p) = Z_\ell(p) \frac{1}{p^2 - m^2 + i0} ,$$

with $Z_\ell$ a polynomial. Suppose the graph has n vertices and L lines; then the corresponding Feynman amplitude $\mathcal{T}(G)$ is the function of n 4-vectors $x_1, \ldots, x_n$ given by

$$\mathcal{T}(G)(x_1, \ldots, x_n) = \prod_{\ell=1}^{L} \Delta^{(\ell)}\left[\sum_{i=1}^{n} e_i^{(\ell)} x_i\right] , \tag{1}$$

with $e_i^{(\ell)}$ the incidence matrix of the graph (Definition A.6). (We lean throughout on the mathematical props of graph theory and the theory of distributions. The basic definitions we will use, and some simple results we will need, are given in Appendices A and B.)

Feynman amplitudes arise when one studies quantum field theory by perturbation-theoretic techniques [see, for example, 2, 32]. In such a study the amplitudes (whose squares give the probabilities) for, say, a scattering process are given as an infinite sum of Feynman amplitudes for different graphs. Now the validity of these techniques is questionable, and the resulting series is not known to converge in any case and known to diverge in some [20]. Nevertheless, we are interested in studying the terms of the series for several reasons: (a) we would like to prove convergence (or even divergence) for series whose behavior is not yet known, (b) the success of the perturbation approach in quantum electrodynamics suggests that we may have an asymptotic series even if not a convergent one, (c) the analytic properties of the summands in the momentum variables suggest analytic properties which may be valid for a true scattering amplitude. Part of the work in this thesis bears on these problems, as we will discuss later.

There is another problem connected with Feynman amplitudes which must be dealt with before any of the above: the product (1) is in fact not even well defined for many graphs, due to the coincidence of the singularities of the factors (or, in momentum space, due to the divergence of the convolution integrals). Dyson [7] has shown how to define these amplitudes appropriately. In this process of *renormalization* the divergences are attributed to unobservable physical effects; they can then be "subtracted out" to produce finite observable results. To state this in a slightly different way: for any graph G in which the product (1) contains divergences, we may define instead a *renormalized Feynman amplitude*; moreover, the renormalization of graphs in a field theory has an acceptable physical interpretation.

We treat these matters in some detail in Chapter I (see also Appendix C, a brief treatment of the theory of free fields). There we review the basic concepts in the Lagrangian theory of interacting fields, the Bogoliubov-Parasiuk definition of renormalization [2, 3, 27], and the mathematically precise work of Hepp [18] on the properties of this definition (referred to as Bogoliubov-Parasiuk-Hepp renormalization, to distinguish it from the somewhat different approach of Dyson and Salam [7, 29]). The chapter contains a thorough analysis of the implementation of renormalization by counterterms in the Lagrangian, although we do not discuss the physical interpretation. We also give a relatively simple extension of Hepp's results which is needed in Chapter III.

Chapter II defines a new quantity called a *generalized Feynman amplitude*, or GFA for short. This amplitude is *not* associated with a particular Feynman graph, but rather depends on certain parameters $\underline{\lambda}$, $\underline{Q}$, $\underline{e}$, $\underline{Z}$, and $\underline{m}$ (whose specific nature does not concern us here). When these parameters take on appropriate values, the GFA becomes (at least formally) equal to the Feynman amplitudes for various graphs. The "appropriate value" for $\underline{\lambda}$ is a fixed point $\underline{\lambda}^0$; if we take $\underline{\lambda} = \underline{\lambda}^0$, we vary the structure of the graph in question by varying $\underline{Q}$ and $\underline{e}$, and vary the propagators associated with the lines by varying $\underline{Z}$ and $\underline{m}$. That is, considering only the $\underline{Q}$ and $\underline{e}$ dependence, we have gone from the Feynman amplitude, defined for a discrete set (the graphs), to the GFA, which is defined for a continuous variable and is equal to various Feynman amplitudes at discrete points. Stated another way, the GFA is an interpolation of the Feynman amplitude between different graphs.

As implied above, the equality of the Feynman amplitude and the GFA is in some cases only formal. This occurs when the graph in question needs to be renormalized, in which case the Feynman amplitude (1) is not well defined. The parameter $\underline{\lambda}$, which is really an L-tuple $(\lambda_1, \ldots, \lambda_L)$ of complex variables, plays an important role in these divergence problems; this is the subject of Chapter III. Suppose then that $\underline{Q}$, $\underline{e}$, $\underline{Z}$, and $\underline{m}$ have been chosen to correspond to some graph G. The resulting GFA is meromorphic as a function of $\lambda$, and, as stated above, is formally equal to $\mathcal{T}(G)$ at $\underline{\lambda}^0$. Now it turns out that a divergence in the amplitude $\mathcal{T}(G)$, necessitating a renormalization, corresponds precisely to a singularity in the GFA at the point $\underline{\lambda} = \underline{\lambda}^0$. Moreover, we can give a prescription for removing singularities of this type, in such a way that application of the prescription to the GFA produces a correctly renormalized Feynman amplitude for the graph. It is important to note that this prescription does not depend on the structure of the graph (in contrast to the usual renormalization procedure, whose recursive subtractions depend explicitly on various subgraphs, etc.). Thus, for example, we may renormalize a sum of Feynman amplitudes by one operation, rather than treating each summand separately.

In Chapter IV we turn to applications of the dependence of the GFA on the parameters $\underline{Q}$ and $\underline{e}$ (the $\underline{Z}$ and $\underline{m}$ dependence plays only a minor role). We originally introduced these parameters in the hope that we could convert a sum of Feynman amplitudes, such as occurs in the perturbation expansion, into an integral over the parameters $\underline{Q}$ and $\underline{e}$. Such an integral representation might be useful in estimating the size of the sum, for study of the convergence properties of the expansion, or in determining the analytic properties of the sum in the momenta. This program has been only partially successful: we have found such an integral representation, but it is valid only for $\underline{\lambda}$ restricted to a region which does not contain the physical point $\underline{\lambda}^0$. Any applications depend on finding an explicit analytic continuation of the integral to a neighborhood of $\underline{\lambda}^0$.

The following notation will be used throughout; it is all standard except possibly (b). Further notation is introduced where appropriate; see in particular Appendices A and B.

(a) If x and y are 4-vectors, we write

$$x \cdot y = \sum_{\mu,\nu=0}^{3} x^{\mu} g_{\mu\nu} y^{\nu} \quad ,$$

where

$$g_{\mu\nu} = \begin{cases} 1, & \text{if } \mu = \nu = 0, \\ -1, & \text{if } \mu = \nu = i > 0, \\ 0, & \text{if otherwise.} \end{cases}$$

(b). In general, we denote a k-tuple of variables $(x_1, \ldots, x_k)$ by underlining: $(x_1, \ldots, x_k) = \underline{x}$. We will use this notation with different values of k simultaneously [e.g., $(a_1, \ldots, a_L) = \underline{a}$, $(c_1, \ldots, c_p) = \underline{c}$]; the dimension of each variable will be clear from context.

We also write:

(c). $x_1, \ldots, \hat{x}_i, \ldots, x_n$ – the variable $x_i$ is omitted;

(d). $[x]$ = the greatest integer less than or equal to the real number x;

(e). $\#(K)$ = the number of elements in the finite set K;

(f). $S_n$ = the complete permutation group on n elements;

(g). det A = the determinant of the matrix A;

(h). $A(^{i}_{j})$ = the (signed) cofactor of the entry $A_{ij}$ in the matrix A;

(i). $A(^{i\,j}_{i\,k})$ = the (signed) cofactor of $A_{jk}$ in the matrix obtained by deleting the $i^{\text{th}}$ row and column of A. $A(^{i\,j}_{i\,k})$ vanishes if i = j or i = k.

We comment finally on the organization of the thesis. Each chapter is divided into sections numbered 1, 2, ... etc.; these are sometimes further subdivided for clarity, using the sequence capital letter, small Roman numeral, small letter [e.g., there is a Chapter I, Section 3 (c. iii. a)]. Theorems, definitions, remarks, etc., are numbered in one continuous sequence for each chapter (e.g., Definition 1.1, etc.) and appendix (Definition A.1). Equations are numbered similarly, but in different sequences. All references are to the bibliography at the end of the thesis.

# CHAPTER I

## *Renormalization in Lagrangian Field Theory*

### Section 1. INTRODUCTION

This chapter is devoted to a brief discussion of Lagrangian Quantum Field Theory, the subject which underlies the entire thesis but which will be mentioned explicitly only in this chapter. The review of the basic material (Section 2) is much too brief even to be called self-contained; we are interested mainly in motivating the study of the perturbation series for the time-ordered vacuum expectation values of a theory and of the Feynman integrals which are the terms of this series. These integrals are often divergent, and renormalization theory is the study of ways to modify them to convergent integrals while maintaining an acceptable physical interpretation of the theory. The final formulas for these "renormalized" integrals have a complicated combinatorial structure. In Section 3 we discuss the relation of these formulas to the field theory, given by the idea of "counterterms" in the Lagrangian of the theory. Finally, Section 4 summarizes the mathematically rigorous work of Hepp on the properties of these renormalized amplitudes, and gives some extensions of his results which we will need in Chapter III.

In this chapter we will frequently discuss products of non-commuting operators on Hilbert space. We therefore adopt the following convention: if $\{A_i \mid i = 1,\ldots,n\}$ are such operators, then

$$\prod_{i=1}^{n} A_i = A_1 A_2 \cdots A_n \; .$$

Similarly,

$$\prod_{i=1}^{n} \prod_{j=1}^{m(j)} A_{ij} = A_{11} \cdots A_{1m(1)} A_{21} \cdots A_{n,m(n)} \; .$$

### Section 2. FIELD THEORY

Our discussion of field theory will follow the notation and spirit of *PCT, Spin and Statistics, and All That,* by Streater and Wightman [34]. Thus the physical states of the theory form a Hilbert space $\mathcal{H}$, which is equipped with a unitary representation,

$$\{a, A\} \to U(a, A) \;,$$

of the group "inhomogeneous $SL(2, \mathbb{C})$." A field $\phi$ is, technically speaking, an operator-valued distribution on $\mathcal{S}(R^4)$, but we follow the usual procedure and write symbolically $\phi(x)$, where

$$\phi(f) = \int_{R^4} \phi(x) f(x)\, dx$$

for any $f \in \mathcal{S}(R^4)$. We refer to Streater and Wightman for a discussion of the finite-dimensional representations of SL(2, C); see also [11].

(A). The free field of mass m.

(i). Transformation law.

Our free field $\phi$ actually consists of M components $\{\phi_1, \ldots, \phi_M\} = \{\phi_a\}$. They transform under Lorentz transformations by the formula

$$U(a, A)\phi_\alpha(x)\, U(a, A)^{-1} = \sum_{\beta=1}^{M} S_{\alpha\beta}(A^{-1})\phi_\beta(Ax+a) \,, \tag{1.1}$$

where S(A) is some M-dimensional representation of SL(2, C). (Note that $A \in SL(2, C)$ corresponds to some Lorentz transformation $\Lambda(A)$, and we abbreviate $\Lambda(A)x$ by $Ax$.) S is not assumed to be irreducible, but must satisfy two conditions:

(a). For all representations $\mathfrak{D}^{(j/2,\, k/2)}$ [see 34] occuring in S, the sum $j+k$ has the same parity;

(b). For any integers j, k, the representations $\mathfrak{D}^{(j/2,\, k/2)}$ and $\mathfrak{D}^{(k/2,\, j/2)}$ occur the same number of times in S.

Now the representation $\mathfrak{D}^{(j/2,\, k/2)}$ is a spinor (two-valued) or tensor (single-valued) representation of the Lorentz group when $j+k$ is odd or even, respectively. Thus (a) enables us to speak of $\phi$ as a half-integral or integral spin field, respectively. By the well-known connection of spin and statistics, then, (a) implies that $\phi$ is either a fermion or boson field. Condition (b) implies the existence of an $M \times M$ invertible Hermitian matrix $\eta$ satisfying

$$\eta^{-1} S(A)^\dagger \eta = S^{-1}(A) \tag{1.2}$$

for any $A \in SL(2, C)$ (here $S(A)^\dagger$ is the Hermitian adjoint of the matrix S(A)) [11]. Equations (1.1) and (1.2) imply that the quantity $\Sigma_{\alpha,\beta}\, \phi_\alpha^*(x)\, \eta_{\alpha\beta}\, \phi_\beta(x)$ transforms like a scalar field under Lorentz transformations, so that

$$\int d^4x \sum_{\alpha,\beta} \phi_\alpha^*(x)\, \eta_{\alpha\beta}\, \phi_\beta(x) \tag{1.3}$$

is a scalar. We will use this matrix $\eta$ in constructing a Lagrangian for the free field. Note that if we consider $\phi$ to be a column vector,

$$\phi = \begin{bmatrix} \phi_1 \\ \vdots \\ \phi_M \end{bmatrix} ,$$

and let $\phi^\dagger$ be the row vector $[\phi_1^*, \ldots, \phi_M^*]$, then (1.3) may be written

$$\int d^4x\, \phi^\dagger(x)\, \eta\, \phi(x) \,.$$

(ii). Field equation.

We assume that $\phi$ satisfies a first-order differential equation of the form

$$(1.4) \qquad \left(-i\gamma^\mu\left(\frac{\partial}{\partial x^\mu}\right)+m\right)\phi(x) = 0 .$$

Here $\gamma^0, \gamma^1, \gamma^2, \gamma^3$ are $M\times M$ matrices which satisfy

$$(1.5) \qquad \eta^{-1}(\gamma^\mu)^\dagger \eta = \gamma^\mu .$$

(A field $\psi$ which satisfies an equation of higher order may be treated in this format by including partial derivatives of $\psi$ among the components of $\phi$; this is one of the reasons we did not require irreducibility of the representation $S$.) The equation (1.4) must be relativistically invariant; that is, if $\phi(x)$ satisfies the equation, so must $U(a, A)\phi(x)U(a, A)^{-1}$, for any $\{a, A\}$ in inhomogeneous $SL(2, C)$. This requires that the $\gamma^\mu$ satisfy

$$(1.6) \qquad S(A)^{-1}\gamma^\mu S(A) = \Lambda(A)^\mu{}_\nu \gamma^\nu$$

for any $A \in SL(2, C)$.

The Lagrangian density for the field may be taken to be

$$\mathcal{L}(x) = \frac{1}{2i}[\phi(x)^\dagger \eta\gamma^\mu \partial_\mu \phi(x) - \partial_\mu \phi(x)^\dagger \eta\gamma^\mu \phi(x)] + m[\phi(x)^\dagger \eta\phi(x)] .$$

This means that equation (1.4) may be derived from $\mathcal{L}(x)$ by the usual principle of least action:

$$\delta \int \mathcal{L}(x)\,dx = 0 .$$

Equation (1.5) is needed in this derivation. Note that (1.2) and (1.7) guarantee that $\mathcal{L}(x)$ is a scalar field:

$$U(a, A)\mathcal{L}(x)U(a, A)^{-1} = \mathcal{L}(Ax+a) .$$

Finally, the requirement that $\phi$ be a field of mass $m$ means that each component of $\phi$ must satisfy the equation

$$\left[\frac{\partial}{\partial x^\mu}\frac{\partial}{\partial x_\mu} - m^2\right]\phi_\alpha(x) = 0 .$$

This is guaranteed by requiring that the only non-zero eigenvalues of $\gamma^0$ be $\pm 1$ [11; see also Appendix C]. *We always take* $m > 0$.

(iii). Quantization

The discussion up to this point could apply also to a classical c-number field $\phi$, if (1.1) were replaced by a transformationlaw

$$\phi'_\alpha(Ax+a) = \sum_\beta S_{\alpha\beta}(A)\phi_\beta(x) .$$

The quantized field $\phi$ may be obtained from this classical field by the usual process of second quantization. A brief discussion of this procedure is given in Appendix C; several additional assumptions about the matrices $\gamma^\mu$ are needed. For our purposes we need three results of this discussion.

(a). Charge conjugation.

The equation (1.4) may be quantized in two essentially different ways: as a self-charge-conjugate (SCC) field or as a non-SCC field. In the second case the particles associated with the field have distinct antiparticles; in the first each particle is its own antiparticle. We are mainly interested in the fact that an SCC field $\phi$ satisfies a relation of the form

$$\phi_\alpha^*(x) = \sum_\beta C_{\alpha\beta}\phi_\beta(x), \tag{1.8}$$

where C is some invertible matrix.

(b). Commutation relations.

The components of the field satisfy the commutation relations

$$[\phi_\alpha^*(x), \phi_\beta(y)]_\pm = Z_{\alpha\beta}\left(\frac{\partial}{\partial x}\right)\Delta(x-y; m) . \tag{1.9}$$

Here the + (–) sign is taken when $\phi$ is fermion (boson) field. $Z_{\alpha\beta}$ is a polynomial, and

$$\Delta(x-y; m) = -\frac{i}{(2\pi)^3}\int d^4k\, e^{-ik\cdot x}\,\delta(k^2-m^2)\,\epsilon(k) .$$

If $\phi$ is non-SCC, we also have

$$[\phi_\alpha(x), \phi_\beta(y)]_\pm = 0 ; \tag{1.10}$$

the commutation relations of an SCC field with itself follow from (1.8) and (1.9).

(c). Normal product

The Lagrangian (1.7) is actually not correct for a quantized field. To avoid infinite energies we must Wick order it:

$$\mathcal{L}(x) = \frac{1}{2i} : [\phi(x)^\dagger\eta\gamma^\mu\partial_\mu\phi(x) - \partial_\mu\phi(x)^\dagger\eta\gamma^\mu\phi(x)] : + m : [\phi(x)^\dagger\eta\phi(x)] : \tag{1.11}$$

(Recall that any free field $\phi$ is a sum $\phi = \phi_1 + \phi_2$, with $\phi_1$ a creation operator and $\phi_2$ a destruction operator. The *normal* or *Wick product* of a family of such fields is defined by

$$:\prod_{i=1}^n \phi^{(i)}: = \sum_B \sigma_B \prod_{i\in B}\phi_1^{(i)} \prod_{i\notin B}\phi_2^{(i)}, \tag{1.12}$$

where the sum is over all subsets $B \subset \{1 \cdots n\}$, and $\sigma_B$ is the sign of the permutation of fermion fields involved in passing from the left hand side of (1.12) to the right hand side.)

(B). Several free fields.

If we have several free fields $\phi^{(1)} \cdots \phi^{(I)}$, we may construct a field theory involving all of them by taking the tensor product of the Hilbert spaces for each; the operators U(a, A) and the fields $\phi^{(i)}(x)$ are defined on the tensor product in a natural way. Thus $\phi^{(i)}$ has $M_i$ components and satisfies

(1.13) $$U(a, A)\phi^{(i)}(x)\,U(a, A)^{-1} = S^{(i)}(A^{-1})\phi^{(i)}(Ax+a)$$

and

(1.14) $$(-i\gamma^{(i)\mu}\frac{\partial}{\partial x^{\mu}} + m_i)\,\phi^{(i)}(x) = 0 .$$

The field equations (1.14) may be derived from the Lagrangian density

$$\mathcal{L}_0(x) = \sum_{i=1}^{I} \mathcal{L}_i(x)$$

(1.15) $$= \frac{1}{2i}\sum_{i=1}^{I} \{ :(\phi^{(i)}(x)^{\dagger}\eta^{(i)}\gamma^{(i)\mu}\partial_{\mu}\phi^{(i)}_{(x)} - \partial_{\mu}\phi^{(i)}(x)^{\dagger}\eta^{(i)}\gamma^{(i)\mu}\phi^{(i)}(x)): + m :(\phi^{(i)}(x)^{\dagger}\,\eta^{(i)}\phi^{(i)}(x)): \} .$$

The fields just defined will satisfy commutation relations

(1.16) $$[\phi^{(i)*}_{\alpha}(x)\, ,\, \phi^{(j)}_{\beta}(y)]_{\pm} = \delta_{ij}\, Z^{(i)}_{\alpha\beta}\,(\frac{\partial}{\partial x})\,\Delta(x-y;\, m_i)\, ,$$

(1.17) $$[\phi^{(i)}_{\alpha}(x),\, \phi^{(j)}_{\beta}(y)]_{\pm} = \delta_{ij} \times \begin{cases} 0, & \text{for } \phi^{(i)} \text{ non-SCC;} \\ \sum_{\delta} C^{(i)-1}_{\alpha\delta}\,[\phi^{(i)*}_{\delta}(x),\, \phi^{(j)}_{\beta}(y)]_{\pm}\, , & \text{for } \phi^{(i)*} = C^{(i)}\phi\, . \end{cases}$$

Here, the + sign holds when $i = j$ and $\phi^{(i)}$ is a fermion field, otherwise, the − sign is taken. However, it is possible to redefine the fields by a Klein transformation [34] so that (1.13), (1.14), (1.16) and (1.17) still hold, but so that the sign in (1.16) and (1.17) is + whenever $\phi^{(i)}$ and $\phi^{(j)}$ are both fermion fields, − otherwise. These are the "normal" commutation relations. We will always assume that this redefinition has been made.

(C). Interacting fields.

(i). Preliminaries.

We now study the theory of a new set of fields $\Phi^{(1)} \cdots \Phi^{(I)}$. These fields correspond directly to the free fields discussed in (B), in particular, they satisfy the transformation law (1.13), with $\Phi^{(i)}$ replacing $\phi^{(i)}$. Their commutation relations are

$$[\Phi^{(i)}(x),\, \Phi^{(j)}(y)]_{\pm} = [\Phi^{(i)*}(x),\, \Phi^{(j)}(y)]_{\pm} = 0$$

whenever $(x-y)^2 < 0$; the + sign holding when both are fermion fields, the − sign otherwise. They satisfy a new equation of motion, however, which is derived in the usual way from the Lagrangian density

$$\mathcal{L}(x) = \mathcal{L}_0(x) + \mathcal{L}_I(x)\; .$$

Here $\mathcal{L}_0(x)$ is given by (1.15) (with $\phi^{(i)}$ replaced by $\Phi^{(i)}$) and $\mathcal{L}_I$ is the interaction Lagrangiar

We will consider quite general interactions. Thus let $\Xi^{(i)}$ be the field whose components are all the components of $\Phi^{(i)}$ together with their derivatives up to some finite but unspecified order. Then $\mathcal{L}_I$ has the form

$$\mathcal{L}_I(x) = \sum_{q=1}^{Q} \mathcal{L}_I^{(q)}(x) = \sum_{q=1}^{Q} g \left\{ \sum_{\underline{a}} M^{(q)}_{a_1 \cdots a_{s(q)}} : \prod_{i=1}^{s(q)} \Psi^{(q,i)}_{a_i}(x) : \right\}$$

where $\Psi^{(q,k)}$ is one of the fields $\Xi^{(i)}$ or, if $\Phi^{(i)}$ is non-SCC, $\Xi^{(i)*}$. $M^{(q)}_{a_1 \cdots a_{s(q)}}$, is some complex coefficient, and g is the (real, positive) interaction strength. Thus $\mathcal{L}_I^{(i)}$ describes the interaction of s(q) fields, possibly with derivative coupling.

The interaction Lagrangian must satisfy several conditions to be physically acceptable.

(a). $\mathcal{L}_I(x)$ must transform as a scalar field under Lorentz transformations, i.e.,

(1.19) $$U(a, A)\, \mathcal{L}_I(x)\, U(a, A)^{-1} = \mathcal{L}_I(Ax + a) .$$

This puts certain restrictions on the $M^{(q)}$, which may be derived from (1.13). It also implies that, for each q, the set $\{\Psi^{(q,i)}\}_{i=1}^{s(q)}$ contain an even number of fermion fields; this follows from (1.19) by taking $a = 0$, $A = \begin{bmatrix} -1 & 0 \\ 0 & -1 \end{bmatrix}$.

(b). $\mathcal{L}_I(x)$ must be Hermitian: $\mathcal{L}_I^*(x) = \mathcal{L}_I(x)$.

(c). $\mathcal{L}_I(x)$ must satisfy "gauge invariance of the first kind" which guarantees the conservation of charge. This means that $\mathcal{L}_I(x)$ is to be invariant when all non-SCC fields $\Phi^{(i)}$ are modified by

$$\Phi^{(i)}(x) \to e^{i\alpha e_i}\, \Phi^{(i)}(x) ,$$

$$\Phi^{(i)*}(x) \to e^{-i\alpha e_i}\, \Phi^{(i)*}(x) ,$$

for any $\alpha$. Here $e_i$ is the charge of the particle associated with $\Phi^{(i)}$.

A cautionary word is necessary. All this discussion has been quite formal, and no such theory is known to exist. Moreover, we will shortly utilize results derived from the interaction picture for this field theory, and this picture definitely does not exist in general [26, 34]. But there seem to be good reasons for studying the theory in this questionable fashion. We will discuss this point more fully below.

(ii). Time-ordered vacuum expectation values.

The basic quantities of this field theory which we will study are related to the time-ordered vacuum expectation values (TOVEV's). A typical TOVEV is denoted by

(1.20) $$(\Omega, T(\Psi^{(1)}(x_1) \cdots \Psi^{(R)}(x_R)\Omega)) .$$

Here $\Omega$ is the vacuum state of the field theory and T is Wick's time ordering operator, defined by

(1.21 $$T\left( \prod_{i=1}^{n} \Psi^{(i)}(x_i) \right) = \sigma \prod_{i=1}^{n} \Psi^{(j_i)}(x_{j_i})$$

whenever $x_{j_1}^0 \geq x_{j_2}^0 \geq \cdots \geq x_{j_n}^0$. In (1.21), $\sigma$ is the parity of the permutation of the fermion operators necessary to pass from the left hand side of the equation to the right hand side. The fact that the fields commute or anticommute for space-like separated arguments implies that (1.21) is well defined except when $x_i = x_j$ for some $i \neq j$. The difficulty in this case is the subject of renormalization theory, which we discuss below. We remark that the S-matrix of the theory is given by the TOVEV's through the reduction formulae of Lehman, Symanzik, and Zimmerman [27]. By using the interaction picture Gell-Man and Low [14] derive an expression for the TOVEV's of these interacting fields in terms of quantities associated with the system of free fields discussed in (B). Their result is

$$(1.22)\qquad (\Omega, T[\Psi^{(1)}(x_1)\cdots \Psi^{(R)}(x_R)]\Omega) = \frac{(\omega, T[U\psi^{(1)}(x_1)\cdots \psi^{(R)}(x_R)]\omega)}{(\omega, U\omega)},$$

where $\omega$ is the vacuum state of the free field theory, and $\psi^{(i)}$ is the free field corresponding to the interacting field $\Psi^{(i)}$. The operator U is usually given as the power series in the coupling constant g:

$$(1.23)\qquad U = \sum_{m=0}^{\infty} \frac{(-i)^m}{m!} \int dy_1 \cdots \int dy_m \, T[\mathcal{L}_I'(y_1)\cdots \mathcal{L}_I'(y_m)] .$$

The $\mathcal{L}_I'$ occurring in (1.23) is the interaction Lagrangian (1.18) with the fields $\Phi^{(i)}$ replaced by $\phi^{(i)}$. If one now substitutes (1.23) into (1.22) and applies Wick's theorem [39] to the time-ordered products, one derives a formula for the TOVEV as a ratio of two power series.

Now we will not, in fact, work directly with the TOVEV's. This is because they are not necessarily covariant; this is true even for force fields in which there is no problem defining the TOVEV's [38]. We will instead deal with covariant versions, which may differ from the TOVEV's by distributions concentrated on subspaces where various $x_i$ coincide. We denote such a *covariant TOVEV* by replacing the T operator in (1.20) by an operator $P^*$:

$$(1.23a)\qquad (\Omega, P^*[\Psi^{(1)}(x_1)\cdots \Psi^{(R)}(x_R)]\Omega) .$$

(This notation comes from the similar notation of [35, 36].) Since we work only in perturbation theory, we will be content to define (1.23) by its perturbation series, given in (1.26) below.

The non-covariance of the T product also leads to difficulties in the right hand sides of (1.22) and (1.23). For a discussion of the corresponding result for the S-matrix (where the difficulties in the definition of the T product of interacting fields does not occur), which includes a careful treatment of the interaction picture, see [36]. Here we simply accept the fact that the T-products on the right hand sides of (1.22) and (1.23) must be replaced by covariant $P^*$ products. The $P^*$ product is defined explicitly for free fields as follows:

*Definition 1.1:* Recall the definition (p.10) of the fields $\Xi^{(i)}$; we define the fields $\xi^{(i)}$ similarly, except that they are obtained by taking derivatives of the interaction-picture fields $\phi^{(i)}$ rather than the Heisenberg fields $\Phi^{(i)}$. The (anti)-commutation relations of the $\xi^{(i)}$ with each other and with their adjoints follow directly from (1.16) and (1.17). Suppose $\psi^{(1)}$ and $\psi^{(2)}$ are two such fields; their (anti)-commutator may be written in the form

$$\left[\psi^{(1)}_{\beta}(y),\ \psi^{(2)}_{\alpha}(x)\right]_{\pm} = Z_{\alpha\beta}\left(\frac{\partial}{\partial x}\right)\Delta(x-y;\ m)$$

where $Z_{\alpha\beta}$ is a polynomial. (We use the anti-commutator for two fermion fields, otherwise the commutator.) Then the *contraction* of the fields $\psi^{(1)}$ and $\psi^{(2)}$ is defined to be

$$\overbrace{\psi^{(1)}_{\beta}(y)\,\psi^{(2)}_{\alpha}(x)} = i\,Z_{\alpha\beta}\left(\frac{\partial}{\partial x}\right)\Delta_F(x-y;\ m)\,,$$

where $\Delta_F$ is the Feynman propagator

$$\Delta_F(x-y;\ m) = \lim_{\epsilon\to 0}\frac{i}{(2\pi)^4}\int d^4k\,\frac{e^{-ik\cdot x}}{k^2-m^2+i\epsilon} \tag{1.24}$$

(Note that our $\Delta_F$ differs by a factor of 2 from that usually used.)

*Definition 1.2:* Let $\chi^{(1)},\ldots,\chi^{(P)}$ be chosen from among the fields $\xi^{(i)}$ (and $\xi^{(i)*}$, if $\phi^{(i)}$ is non-SCC). Then, if P(1),..., P(s) satisfy $1 \le P(1) < P(2) < \cdots < P(s) = P$, define

$$P^*(:\chi^{(1)}\cdots\chi^{(P(1))}:\ :\chi^{(P(1)+1)}\cdots:\ :\chi^{(P(s-1)+1)}\cdots\chi^{(P)}:)$$

$$= \sum_{r=0}^{\left[\frac{P}{2}\right]}\ \sum_{\substack{i_1,\ldots i_r\\ j_1,\ldots j_r}} \overbrace{\chi^{(i_1)}\chi^{(j_1)}}\cdots\overbrace{\chi^{(i_r)}\chi^{(j_r)}}\ :\chi^{(1)}\cdots\hat{\chi}^{(i_1)}\cdots\hat{\chi}^{(j_r)}\cdots\chi^{(P)}: \tag{1.25}$$

Here $i_1,\ldots i_r,\ j_1,\ldots j_r$ run over all 2r-tuples of integers satisfying

(a). $1 \le i_k \le P,\quad 1 \le j_k \le P$;

(b). $i_1 < i_2 < \cdots < i_r,\ j_1 < j_2 < \cdots < j_r$;

(c). for each k with $1 \le k \le r$ there exists a $k'$ with $1 \le k' \le s$ satisfying $i_k \le P(k') < j_k$.

In (1.25), $\sigma$ is the sign of the permutation of fermion fields that occurs when passing from the order $(1, 2, \ldots P)$ to the order $(i_1, j_1, \ldots i_r, j_r, 1, 2, \ldots \hat{i}_1, \ldots \hat{j}_r, \ldots P)$.

It is important to recognize the similarity of Definition 1.2 to Wick's theorem. The expansions of the $P^*$ product and the T product are combinatorially the same; the difference is in the definitions of the contractions. The corrected formulas (1.22) and (1.23) are now

$$(\Omega, P^*[\Psi^{(1)}_{\alpha_1}(x_1)\cdots\Psi^{(R)}_{\alpha_R}(x_R)]\,\Omega) = \frac{(\omega, P^*[U\,\psi^{(1)}_{\alpha_1}(x_1)\cdots\psi^{(R)}_{\alpha_R}(x_R)]\,\omega)}{(\omega, U\omega)}\,, \tag{1.26}$$

$$U = \sum_{m=0}^{\infty}\frac{(-i)^m}{m!}\int dy_1\cdots dy_m\,P^*[\mathcal{L}_I'(y_1)\cdots\mathcal{L}_I'(y_m)]\,. \tag{1.27}$$

We emphasize that we look on (1.26) as a definition of its left hand side.

(iii). Graphs.

We now apply Definition 1.2 to the $P^*$ products which occur when (1.27) is substituted into (1.26). The numerator of (1.26) becomes

$$\sum_{m=0}^{\infty} \frac{(-i)^m}{m!} \int dy_1 \cdots dy_m (\omega, P^*[\mathcal{L}_I(y_1) \cdots \mathcal{L}_I(y_m) \quad \psi^{(1)}_{\alpha_1}(x_1) \cdots \psi^{(R)}_{\alpha_R}(x_R)]\,\omega) \,. \tag{1.28}$$

If we use (1.18), (1.28) becomes

$$\sum_{m=0}^{\infty} \frac{(-ig)^m}{m!} \sum_q \sum_{\underline{\alpha}} \prod_{k=1}^{m} M^{(q(k))}_{\alpha_{1,k} \cdots \alpha_{s(q(k)),k}} \int dx_{R+1} \cdots dx_{R+m}$$
$$\left(\omega, P^* \left\{ \prod_{k=1}^{m} : \left[ \prod_{i=1}^{s(q)} \psi^{(q(k),i)}_{\alpha_{i,k}}(x_{k+R}) \right] : \left[ \prod_{i=1}^{R} \psi^{(i)}_{\alpha_i}(x_i) \right] \right\} \omega \right) \tag{1.29}$$

Here $\Sigma_q$ is over all maps $q: \{1,\ldots,m\} \to \{1,\ldots,Q\}$. Now we apply (1.25) to the scalar product in (1.29), recalling that the vacuum expectation value of any normal product vanishes. Thus the scalar product vanishes unless $[\Sigma_{k=1}^m s(q(k)) + R]$ is even, in which case it reduces to the sum over all ways of completely contracting the $P^*$ product.

These complete contractions are related to Feynman graphs as follows. The graph involved has $R+m$ vertices, $V_1, \ldots, V_{R+m}$. For each pair of contracted fields, say $\overline{\chi_\beta(x_i)\chi'_\gamma(x_j)}$, the graph contains a line $\ell$ directed from $V_i = V_{i_\ell}$ to $V_j = V_{f_\ell}$; and with this line we associate a "propagator"

$$\Delta^{(\ell)}_{\gamma\beta}(x_{f_\ell} - x_{i_\ell}) = \overline{\chi_\beta(x_i)\,\chi'_\gamma(x_j)} \tag{1.30}$$

If for convenience we write $\alpha_1(\ell) = \beta$, $\alpha_2(\ell) = \gamma$, then (1.29) becomes

$$\sum_{m=0}^{\infty} \frac{-ig)^m}{m!} \sum_q \sum_{\underline{\alpha}} \prod_{k=1}^{m} M^{q(k)}_{\alpha_{1,k} \cdots \alpha_{s(q(k)),k}} \int dx_{R+1} \cdots dx_{R+m}$$
$$\Sigma' \sigma \prod_{\mathcal{L}} \Delta^{(\ell)}_{\alpha_2(\ell)\,\alpha_1(\ell)}(x_{f_\ell} - x_{i_\ell}) \tag{1.31}$$

Here $\Sigma'$ is the sum over all ways of contracting the $P^*$ product in (1.29), or equivalently the sum over all Feynman graphs, $\sigma$ is the sign factor from (1.25), and $\mathcal{L}$ is the collection of lines of the graph. We will give a more precise description of the graphs which can occur in Section 3.D.

The denominator of (1.26) may be handled similarly. In this case, the graphs of $m^{th}$ order have only the m vertices corresponding to $\mathcal{L}_I(y_1) \cdots \mathcal{L}_I(y_m)$. Such a graph is called a vacuum-vacuum graph. Now any graph in the sum (1.31) may be decomposed into components. If we group together all graphs whose structure differs only by vacuum-vacuum components, we may see that (1.31) actually contains a factor $(\omega, U\omega)$ which cancels the denominator of (1.26). Thus we finally have

$$(\Omega, P^*[\Psi^{(1)}_{a_1}(x_1)\cdots\Psi^{(R)}_{a_R}(x_R)]\Omega) = \sum_{m=0}^{\infty}\frac{(-ig)^m}{m!}\sum_q{\sum}''\sum_{\underline{a}} \tag{1.32}$$
$$\prod_{k=1}^{m} M^{(q(k))}_{a_{1,k}\cdots a_{s(q(k)),k}} \int dx_{R+1}\cdots dx_{R+m}\,\sigma \prod_{\mathcal{L}} \Delta^{(\ell)}_{a_2(\ell)a_1(\ell)}(x_{f_\ell}-x_{i_\ell}).$$

Here $\Sigma''$ is over all graphs with no vacuum-vacuum components.

We will actually find it more convenient to study the *truncated covariant* TOVEV's [1, 17] defined recursively by

$$(\Omega, P^*[\Psi^{(1)}(x_1)\cdots\Psi^{(R)}(x_R)]\Omega) = \tag{1.33}$$
$$\sum \sigma \prod_{j=1}^{j(P)} (\Omega, P^*[\Psi^{(i(1,j))}(x_{i(1,j)})\cdots\Psi^{(i(r(j),j))}(x_{i(r(j),j)})]\Omega)^T .$$

Here the sum is taken over all partitions P of $\{1,\ldots,R\}$ into $j(P)$ non-empty subsets $\{i(1,j),\ldots$ $i(r(j),j)\}$, with $i(1,j) < i(2,j) < \cdots < i(r(j),j)$, and $\sigma$ is the usual sign factor associated with the change in order of the fermion fields. It is easily verified from (1.32) and (1.33) that the truncated covariant TOVEV

$$(\Omega, P^*[\Psi^{(1)}_{a_1}(x_1)\cdots\Psi^{R}_{a_R}(x_R)]\Omega)^T$$

is given by the right hand side of (1.32) with $\Sigma''$ replaced by $\Sigma'''$, the sum over all *connected* graphs. It is this series for the truncated covariant TOVEV's that we will study in the next section.

(iv). Justification.

We now return to a question raised earlier—the justification of all these manipulations. In addition to the questions raised about the existence of (a) the field theory and (b) the interaction picture, we now have two more problems: (c) we have produced an infinite series whose convergence is highly questionable (it is known to diverge in some cases), and (d) the terms of this series are not well defined, since the product of $\Delta_F$ functions contains "ultraviolet divergences."

It is problem (d) that will concern us for the rest of this chapter. The method for multiplying propagators is called renormalization theory; it was developed first by Dyson [7], but we shall follow more closely the ideas of Bogoliubov, Parasiuk, and Hepp [2, 3, 18, 27]. There are several reasons for spending time on this problem. First, there is the remarkable success of the theory in quantum electrodynamics: once the series (1.32) has been renormalized, the first few terms approximate physical reality very closely. Thus, in this case at least, (1.32) is probably valid as an asymptotic series. Secondly, if the renormalized version of (1.32) could be shown to be convergent in some case, the result might well be valid TOVEV's for a field theory despite the questionable derivation. Finally, the analytic properties of the terms of (1.32) (or the terms of the renormalized series) in the momentum variables dual to $x_1, \ldots, x_R$ serve to indicate the analytic properties we may expect for the TOVEV's of a real field theory.

Section 3. RENORMALIZATION

(A). Preliminaries.

We now turn to the question of defining the product of propagators appearing in (1.32). Let us restate the problem in more general terms. We have a graph G, with vertices $V_1 \cdots V_n$, and a set $\mathcal{L}$ of (oriented) lines. For each line $\ell \in \mathcal{L}$ we have a propagator $\Delta^{(\ell)}(x)$, a distribution in $\mathcal{S}'(R^4)$, given in p-space by

$$\tilde{\Delta}^{(\ell)}(p) = Z^{(\ell)}(p) \frac{i}{(2\pi)^2} \frac{1}{p^2 - m_\ell^2 + i0}, \tag{1.33}$$

where $m_\ell > 0$ and $Z^{(\ell)}$ is a polynomial of degree $r_\ell$. Let $V_{i_\ell}$ and $V_{f_\ell}$ be the initial and final vertices of the line $\ell$. Then we wish to define a distribution $\mathcal{T}(G) \in \mathcal{S}'(R^{4m})$ (called the *Feynman amplitude* of the graph) given in some sense by

$$\mathcal{T}(G) = \prod_{\mathcal{L}} \Delta^{(\ell)}(x_{f_\ell} - x_{i_\ell}) . \tag{1.34}$$

Moreover, our definition is to be such that, when it is applied to the set of graphs arising from a particular field theory, there is to be some consistent physical interpretation of the results.

The difficulty in defining (1.34) arises from the fact that $\Delta_F(x)$ has a strong singularity at $x = 0$, and hence the factors in the product in (1.34) have singularities on those surfaces where two or more $x_i$ coincide. There is no general formula for multiplying such distributions. This difficulty shows up in another way if we consider the Fourier transform of (1.34). Then the product of distributions becomes convolution, and these convolutions give integrals which diverge for large momenta.

We note that, according to (1.32), even after (1.34) has been defined it will have to be integrated over some of the x variables. This turns out to be possible, due to our assumption $m_\ell > 0$ [18]. We will return to this point in Chapter III.

To put our discussion of (1.34) on a precise basis, we introducc regularized propagators $\Delta^{(\ell)}_{\epsilon,r}$ given by modifying (1.33) into

$$\tilde{\Delta}^{(\ell)}_{\epsilon,r} = \frac{Z^{(\ell)}(p)}{(2\pi)^2} \int_r^\infty d\alpha_\ell \, e^{i\alpha_\ell(p^2 - m_\ell^2 + i\epsilon)} \tag{1.35}$$

where $r > 0$, $\epsilon > 0$. Then $\Delta^{(\ell)} = \lim_{\epsilon\to 0+} \lim_{r\to 0+} \Delta^{(\ell)}_{\epsilon,r}$. $\Delta^{(\ell)}_{\epsilon,r}$ belongs to $\mathcal{O}_M(R^4)$ [31, p. 243], the space of polynomially bounded $C^\infty$ functions on $R^4$. Thus the regularized amplitude

$$\mathcal{T}_{\epsilon,r}(G)(x) = \prod_{\mathcal{L}} \Delta^{(\ell)}_{\epsilon,r}(x_{f_\ell} - x_{i_\ell}) \tag{1.36}$$

is a well defined distribution.

At this stage we might attempt to define (1.34) by $\mathcal{T}(G) = \lim_{\epsilon\to 0+} \lim_{r\to 0+} \mathcal{T}_{\epsilon,r}(G)$, but our original difficulty is now reflected in the fact that $\lim_{r \to 0+} \mathcal{T}_{\epsilon,r}(G)$ does not exist. Our program instead is as follows: to perform certain manipulations on $\mathcal{T}_{\epsilon,r}(G)$ which will transform it to a new distribution $\mathcal{R}_{\epsilon,r}(G)$ such that $\mathcal{R}(G) = \lim_{\epsilon \to 0+} \lim_{r \to 0+} \mathcal{R}_{\epsilon,r}(G)$ exists. $\mathcal{R}(G)$ is the *renormalized Feynman amplitude* for the graph G. We remark that the real problem is the $r \to 0+$ limit; once its existence has been achieved, the $\epsilon \to 0+$ limit exists also.

(B). Divergence of Feynman integrals.

The basic idea used in the "manipulations" referred to above is the subtraction procedure of Dyson. Very roughly speaking, one writes the regularized amplitude (1.36) in momentum space and performs a Taylor expansion in the momentum variables. The divergences are contained in the low order terms of this expansion, and if one subtracts these terms, the remainder has a finite $r \to 0+$ limit. Actually, as we shall see in subsection (C), this procedure must be applied recursively, first to the amplitudes of various subgraphs of G, then finally to the amplitude for G itself. We are not concerned with this difficulty here. Rather, we have two goals: to show how subtraction in momentum space can improve the behavior of a Feynman amplitude, and to provide a motivation for the introduction of a number $\mu(G)$, the superficial divergence of the graph G, which tells us how much subtraction we need to do.

Let us first consider a simple example (Figure 1.1). We will suppose that all lines repre-

Figure 1.1

sent spinless particles of mass m, so that

$$\tilde{\Delta}^{(\ell)}(p) = \frac{i}{(2\pi)^2} \frac{1}{(p^2 - m^2 + i0)}$$

for all lines $\ell$. (This graph is in fact a self-energy in the $\phi^4$ theory.) The amplitude $\tilde{\mathcal{T}}_{\epsilon,r}(p_1, p_2)$ is easily calculated [see, e.g., 8; also Remark 3.17] from (1.35) and (1.36); it is given, up to constant factors, by

$$\tilde{\mathcal{T}}_{\epsilon,r}(p_1+p_2) = \delta(p_1+p_2) \int_r^\infty d\alpha_1 \int_r^\infty d\alpha_2 \int_r^\infty d\alpha_3 \times$$

$$\times \frac{1}{(\alpha_1\alpha_2+\alpha_2\alpha_3+\alpha_1\alpha_3)^2} \exp i \left\{ \frac{\alpha_1\alpha_2\alpha_3 p_1^2}{\alpha_1\alpha_2+\alpha_2\alpha_3+\alpha_1\alpha_3} - (m^2 - i\epsilon)(\alpha_1+\alpha_2+\alpha_3) \right\} \tag{1.37}$$

The argument of the exponential is continuous in $\alpha$ and $p_1$ for all $\alpha_\ell \geq 0$, and the integral converges at large values of $\alpha_\ell$ because of the factor $e^{-\epsilon\alpha_\ell}$. The divergence problem arises because the quantity $\alpha_1\alpha_2+\alpha_2\alpha_3+\alpha_1\alpha_3$ vanishes when any two $\alpha$'s vanish; thus, we cannot let $r \to 0$ in (1.37).

To study the nature of the singularity of an integrand like that in (1.37) near a point where certain of the integration variables vanish, say $\alpha_{\ell_1} = \alpha_{\ell_2} = \cdots = \alpha_{\ell_k} = 0$, it is convenient to make the change of variables

$$(1.38)\qquad \begin{aligned} t &= \alpha_{\ell_1} + \cdots + \alpha_{\ell_k}\,,\\ \alpha_{\ell_1} &= t\beta_{\ell_1}\,,\\ \alpha_{\ell_k} &= t\beta_{\ell_k}\,,\end{aligned}$$

where $\Sigma_{i=1}^{k}\beta_{\ell_i} = 1$. We will study only the singularity of (1.37) where all the $\alpha$'s vanish; using (1.38) the integral becomes

$$(1.39)\qquad \begin{aligned} \bar{\mathcal{T}}_{\epsilon,r}(p_1,p_2) &= \delta(p_1+p_2)\int_{3r}^{\infty} dt \int_{r/t}^{\infty} d\beta_1 \int_{r/t}^{\infty} d\beta_2 \int_{r/t}^{\infty} d\beta_2 \,\frac{\delta(1-\sum_1^3 \beta_\ell)\,t^{-2}}{(\beta_1\beta_2+\beta_2\beta_3+\beta_1\beta_3)^2}\\ &\times \exp it\left[\frac{\beta_1\beta_2\beta_3 p_1^2}{\beta_1\beta_2+\beta_2\beta_3+\beta_3\beta_1} - (m^2-i\epsilon)(\beta_1+\beta_2+\beta_3)\right]\end{aligned}$$

Then the divergence of (1.37) at $\alpha_1 = \alpha_2 = \alpha_3 = 0$ corresponds to the divergence t-integral in (1.39).

We now look for a way to remove this divergence. The basic idea of Dyson leads us to the following definition:

*Definition 1.3.* Let $\mu$ be an integer, and let $\mathcal{T} \in S'(R^{4m})$ have the form

$$(1.40)\qquad \mathcal{T}(p_1,\ldots,p_m) = \delta\left(\sum_{i=1}^{m} p_i\right) F(p_1,\ldots,p_m)$$

where F is a $C^\infty$ function. Then the operator $\mathcal{M}_\mu$ is defined by

$$[\mathcal{M}_\mu \mathcal{T}](p_1,\ldots,p_m) = \begin{cases} 0 & \text{if } \mu < 0\,,\\ \delta(\Sigma_1^m p_i)\, G(p_1,\ldots,p_m) & \text{if } \mu \geq 0\end{cases}$$

where G is the Taylor series of F about $p_1 = p_2 = \cdots = 0$ truncated at order $\mu$.

We recall the well known formula ($\mu \geq 0$):

$$(1.41)\qquad F(p_1\cdots p_m) - G(p_1\cdots p_m) = \frac{1}{\mu!}\int_0^1 d\tau(1-\tau)^\mu \frac{\partial}{\partial\tau^{\mu+1}} F(\tau p_1,\ldots,\tau p_m)\,.$$

Now $\mathcal{T}_{\epsilon,r}(p_1,p_2)$ has the form (1.40), so we may define

$$\mathcal{R}_{\epsilon,r}(p_1,p_2) = (1-\mathcal{M}_2)\,\mathcal{T}_{\epsilon,r}(p_1,p_2)$$

and, applying (1.41) to (1.39), we find

(1.42)
$$\mathcal{R}_{\epsilon,r}(p_1,p_2) = -\delta(p_1+p_2)\int_r^\infty d\alpha_1\int_r^\infty d\alpha_2\int_r^\infty d\alpha_3\int_0^1 d\tau[\tau(1-\tau)^2] \times\left\{\frac{\alpha_1\alpha_2\alpha_3 p_1^2}{(\alpha_1\alpha_2+\alpha_2\alpha_3+\alpha_1\alpha_3)^2}\right\}^2\left\{6+4i\,\frac{\tau^2\alpha_1\alpha_2\alpha_3 p_1^2}{\alpha_1\alpha_2+\alpha_2\alpha_3+\alpha_1\alpha_3}\right\} \times\exp i\left\{\frac{\tau^2\alpha_1\alpha_2\alpha_3 p_1^2}{\alpha_1\alpha_2+\alpha_2\alpha_3+\alpha_1\alpha_3}-(m^2-i\epsilon)(\alpha_1+\alpha_2+\alpha_3)\right\}.$$

It is easy to see that the integrand of (1.42) is integrable over the entire region $\alpha_\ell \geq 0$; thus $\lim_{r\to 0+}\mathcal{R}_{\epsilon,r}$ exists. Note that we have obtained $\mathcal{R}_{\epsilon,r}$ by applying the operator $1-\mathfrak{M}_2$ to the Feynman amplitude; that is, by subtracting from the amplitude the terms of its Taylor series up to order two. As we shall see, the superficial divergence for the graph of Fig. 1.1 is two.

We now consider the amplitude $\mathcal{T}_{\epsilon,r}(G)$ for the general graph discussed in (A). We will derive a formula for this amplitude later [Remark 3.17] and find that it has the form

(1.43)
$$\mathcal{T}_{\epsilon,r}(p_1,\ldots,p_n) = \delta\Big(\sum_1^n p_i\Big)\int_r^\infty\cdots\int_r^\infty \prod_{\mathcal{L}} d\alpha_\ell\,\frac{B(\underline{\alpha},\underline{p})}{C(\underline{\alpha})^2}\exp i\left[\frac{D(\underline{\alpha},\underline{p})}{C(\underline{\alpha})}-\sum_{\ell\in\mathcal{L}}\alpha_\ell(m_\ell^2-i\epsilon)\right]$$

Here $C(\underline{\alpha})$ is a polynomial, homogeneous of degree $L-n+1$, $D(\underline{\alpha},\underline{p})$ is a polynomial homogeneous of degree 2 in $\underline{p}$ and of degree $L-n+2$ in $\alpha$, and $B(\underline{\alpha},\underline{p})$ has the form

(1.44)
$$B(\underline{\alpha},\underline{p}) = \sum_{r=0}^{\rho}\sum_{\underline{i},\underline{\mu}}\left\{\sum_{j=0}^{[\frac{\rho-r}{2}]} E_{\underline{i},\underline{\mu},j}(\alpha)\right\} p_{i_1\mu_1}\cdots p_{i_r\mu_r},$$

with $\rho = \sum_{\mathcal{L}} r_\ell$, and with $E_{\underline{i},\underline{\mu},j}(\underline{\alpha})$ a rational function homogeneous of degree $(-j)$. The integral (1.43) diverges when $r\to 0$ because the factors $C(\underline{\alpha})^{-2}$ and $E_{\underline{i},\underline{\mu},j}(\underline{\alpha})$ diverge when various subsets of the $\alpha$'s vanish.

Again we investigate the behavior of the integrand in (1.43) near the point $\alpha_1 = \cdots = \alpha_L = 0$. Making the variable change $\alpha_\ell = t\beta_\ell$ as in (1.38), we find

$$E_{\underline{i},\underline{\mu},\underline{j}}(\alpha) \to t^{-j}E_{\underline{i},\underline{\mu},j}(\underline{\beta}),$$

$$C(\underline{\alpha}) \to t^{L-n+1}C(\underline{\beta}).$$

Thus (1.43) becomes

(1.45)
$$\delta(\Sigma p_i)\int_{nr}^{\infty} dt \int_{r/t}^{\infty} d\beta_1 \cdots \int_{r/t}^{\infty} d\beta_L\, \delta(1-\Sigma\,\beta_\ell) \sum_{r=0}^{\rho} \sum_{\underline{i},\underline{\mu}} \sum_{j=0}^{\left[\frac{\rho-r}{2}\right]}$$
$$\times\; t^{2n-L-j-3}\, E_{\underline{i},\underline{\mu},j}(\underline{\beta})\, C(\underline{\beta})^{-2}\, p_{i_1\mu_1} \cdots p_{i_r\mu_r}$$
$$\times\; \exp it\left[\frac{D(\underline{\beta},\underline{p})}{C(\underline{\beta})} - \sum_{\mathcal{L}} \alpha_\ell(m_\ell^2 - i\epsilon)\right] .$$

Now the power of t in (1.45) satisfies

(1.46)
$$2n - L - j - 3 \geq 2n - L - \frac{\rho}{2} - 3$$
$$= -\frac{1}{2}[2L + \rho - 4(n-1)] - 1 .$$

That is, the t integral converges when $r \to 0$ if $[2L + \rho - 4(n-1)] < 0$. This motivates

*Definition 1.4:* The *superficial divergence of a Feynman* graph G is given by

$$\mu(G) = 2L(G) + \sum_{\ell \in \mathcal{L}(G)} r_\ell - 4[n(G) - 1] .$$

Here L(G) and n(G) are the number of lines and vertices of G, respectively, $\mathcal{L}(G)$ the set of lines of G, and $r_\ell$ the degree of $Z^{(\ell)}$ in (1.33).

From (1.45) and (1.46) we see that $\mu(G)$ measures the degree to which the Feynman integral diverges at $\alpha_1 = \cdots = \alpha_L = 0$.

Finally, consider the distribution $[1-\mathfrak{M}_{\mu(G)}]\,\mathcal{T}_{\epsilon,r}(G)$. Using (1.41), (1.45) and (1.46) we may verify easily that when $[1-\mathfrak{M}_{\mu(G)}]\,\mathcal{T}_{\epsilon,r}(G)$ is written as an $\alpha$-integral and the transformation $\alpha_\ell = t\beta_\ell$ is made, the resulting t-integral is convergent (when $r \to 0$). Thus the "subtraction" $[1-\mathfrak{M}_{\mu(G)}]$ removes the divergence of $\mathcal{T}_{\epsilon,r}(G)$ associated with the point $\alpha_1 = \cdots = \alpha_2 = 0$. Of course, the resulting integral may still be divergent when various subsets of the $\beta$'s vanish. The removal of these divergences requires the recursive subtractions defined in the next subsection.

(C). Renormalization of an arbitrary graph.

In this subsection we give the formulas for the renormalized amplitude $\mathcal{R}_{\epsilon,r}(G)$ for an arbitrary graph G, using the notation of Hepp [18]. In one sense these formulas do not involve any necessary connection with a field theory; they depend only on the structure of the graph itself. It is in this light that we present them here. But if the graph in question arises from an expansion like (1.32), the renormalization has a close connection with the field theory; this will be discussed in the next subsection.

We refer to Appendix A for definitions of a one-particle irreducible (IPI) graph, a one-particle reducible (IPR) graph, and a generalized vertex. If $\{V_1', \ldots, V_m'\} \subset \{V_1, \ldots, V_n\}$ is a generalized vertex and $G' = G(V_1', \ldots, V_m')$ the corresponding subgraph of G, we denote

$\mathcal{T}(G')$, $\mathcal{R}(G')$, etc., by $\mathcal{T}(V_1' \cdots V_m')$, $\mathcal{R}(V_1' \cdots V_m')$, etc. In particular, the regularized Feynman amplitude for the entire graph G is written as $\mathcal{T}_{\epsilon,r}(V_1 \cdots V_n)$.

We can see the importance of the IPI-IPR distinction as follows. Any IPR graph G naturally decomposes into a certain number of IPI components $G^{(1)}, \ldots, G^{(K)}$ and single lines $\ell_1, \ldots, \ell_j$. Let $G^{(k)} = G(V_1^{(k)} \cdots V_{m(k)}^{(k)})$, let $y_i^{(k)} = x_i^{(k)} - x_{m(k)}^k$ $(i = 1,\ldots,m(k)-1)$, and let $Z_i = x_{f_{\ell_i}} - x_{i_{\ell_i}}$ $(i = 1,\ldots,j)$. It follows from the fact that G is IPR that all the variables $y_i^{(k)}$ and $Z_i$ are linearly independent. Now (1.36) may be written

$$\mathcal{T}_{\epsilon,r}(G) = \prod_{k=1}^{K} \mathcal{T}_{\epsilon,r}(G^{(k)}) \prod_{i=1}^{j} \Delta_{\epsilon,r}^{(\ell_i)} \tag{1.47}$$

But $\mathcal{T}_{\epsilon,r}(G^{(k)})$ depends only on $\{y_i^{(k)}\}_{i=1}^{m(k)-1}$, and $\Delta^{(\ell_i)}$ only on $Z_i$; this means that (1.47) is really a direct product of distributions. Thus, once we have defined $\mathcal{R}_{\epsilon,r}(G^{(K)})$ for the IPI graphs $G^{(k)}$, we can define

$$\mathcal{R}_{\epsilon,r}(G) = \prod_{k=1}^{K} \mathcal{R}_{\epsilon,r}(G^{(k)}) \prod_{i=1}^{j} \Delta_{\epsilon,r}^{(\ell_i)} , \tag{1.48}$$

and the existence of the $r \to 0+$ limit in (1.48) will be guaranteed by the existence of that limit for each factor in the direct product.

We now give the basic

*Definition 1.5.* Let G be a Feynman graph, as above. For each generalized vertex $\{V_1' \cdots V_m'\}$ of G define recursively

$$\mathfrak{X}_{\epsilon,r}(V_1',\ldots,V_m') = \begin{cases} 1, & \text{if } m = 1, \\ 0, & \text{if } G(V_1',\ldots,V_m') \text{ IPR}, \\ -\mathfrak{M}_{\mu(V_1',\ldots,V_m')} \bar{\mathcal{R}}_{\epsilon,r}(V_1',\ldots,V_m') & \text{otherwise;} \end{cases} \tag{1.49}$$

$$\bar{\mathcal{R}}_{\epsilon,r}(V_1',\ldots,V_m') = \sum_P \prod_{j=1}^{k(P)} \mathfrak{X}_{\epsilon,r}(V_{j,1}^P \cdots V_{j,r(j)}^P) \prod_{\text{conn}} \Delta_{\epsilon,r}^{(\ell)} ; \tag{1.50}$$

$$\mathcal{R}_{\epsilon,r}(V_1',\ldots,V_m') = \bar{\mathcal{R}}_{\epsilon,r}(V_1',\ldots,V_m') + \mathfrak{X}_{\epsilon,r}(V_1',\ldots,V_m') . \tag{1.51}$$

Here $\Sigma_P$ is over all partitions P of $\{V_1' \cdots V_m'\}$ into $k(P) \geq 2$ disjoint subsets $\{V_{j,1}^P \cdots V_{j,r(j)}^P\}$, and $\Pi_{\text{conn}}$ is the product over all lines of the graph which connect different subsets of the partition.

*Remark 1.6:* The quantity $\mathcal{R}_{\epsilon,r}(V_1,\ldots,V_n)$ is the renormalized amplitude for the graph G. The quantities defined in (1.49)-(1.51) are all distributions in $\mathcal{S}'(\mathcal{R}^{4m})$ (distributions in the variables $x_1',\ldots,x_m'$). $\mathfrak{X}_{\epsilon,r}(V_1',\ldots,V_m')$ is called the *vertex part* (for the generalized vertex $\{V_1',\ldots,V_m'\}$); it is a distribution concentrated on the surface $x_1' = x_2' = \cdots = x_m'$.

Let us examine the structure of (1.49) - (1.51). For the trivial case $m = 1$, we have $\bar{\mathcal{R}}_{\epsilon,r}(V_1') = 0$, $\mathcal{X}_{\epsilon,r}(V_1') = \mathcal{R}_{\epsilon,r}(V_1') = 1$. For $m = 2$, we have

$$\bar{\mathcal{R}}_{\epsilon,r}(V_1', V_2') = \mathcal{T}_{\epsilon,r}(V_1', V_2')$$

and hence

$$\mathcal{R}_{\epsilon,r}(V_1', V_2') = \begin{cases} [1-\mathfrak{M}_{\mu(V_1', V_2')}]\, \mathcal{T}_{\epsilon,r}(V_1', V_2'), \text{ if } G(V_1', V_2') \text{ is IPI}, & (1.52a) \\ \mathcal{T}_{\epsilon,r}(V_1', V_2') \text{ otherwise}. & (1.52b) \end{cases}$$

Case (1.52 a) is just the renormalization we arrived at above for the simple example of Figure 1.1. Case (1.52b) occurs when $G(V_1', V_2')$ contains either no lines ($\mathcal{T}_{\epsilon,r}(V_1', V_2') = 1$) or only one line $\ell$ ($\mathcal{T}_{\epsilon,r}(V_1', V_2') = \Delta_{\epsilon,r}^{(\ell)}$); thus there is no divergence to remove even if $\mu(V_1', V_2') \geq 0$.

Now consider (1.49) - (1.51) for arbitrary $m$. The term in the sum (1.50) for which $k(P) = m$ (and hence $r(j) = 1$ for all $j$) is precisely $\mathcal{T}_{\epsilon,r}(V_1', \ldots, V_m')$. The remaining terms correspond to partitions of $\{V_1', \ldots, V_m'\}$ into at least two and at most $m-1$ subsets. These are subtractions which remove the divergences of the integral (1.43) which are associated with the vanishing of various proper subsets of the $\alpha$'s. Then (1.51) becomes

$$\mathcal{R}_{\epsilon,r}(V_1', \ldots, V_m') = \begin{cases} [1-\mathfrak{M}_{\mu(V_1', \ldots, V_m')}]\, \bar{\mathcal{R}}_{\epsilon,r}(V_1', \ldots, V_m'), \text{ if } G(V_1', \ldots, V_m') \text{ is IPI} & (1.53a) \\ \bar{\mathcal{R}}_{\epsilon,r}(V_1', \ldots, V_m') \text{ otherwise}. & (1.53b) \end{cases}$$

In (1.53a) we make the subtraction discussed in (B), which is sufficient to remove the divergence of (1.43) associated with $\alpha_1 = \cdots = \alpha_L = 0$. Combined with the subtractions already contained in $\bar{\mathcal{R}}_{\epsilon,r}(V_1', \ldots, V_m')$, this produces an amplitude which is finite in the $r \to 0+$ limit [18; see also Section 4]. (1.53b) corresponds to (1.48), which we have already discussed.

Now that we have defined renormalized Feynman amplitudes in general, we show how the subtractions of (1.50) and (1.51) can be implemented in a field theory.

D. Renormalization in a Field Theory.

(i). Introduction.

We can change the Feynman amplitudes occurring in a field theory to the renormalized Feynman amplitudes of Definition 1.5 in a rather simple way. We do this by adding new terms to the interaction Lagrangian, of the form

$$\sum_{\substack{\alpha_1 \ldots \alpha_s \\ \mu_{1,1} \ldots \mu_{s,r(s)}}} M^{\mu_{1,1} \ldots \mu_{s,r(s)}}_{\alpha_1 \ldots \alpha_s} : \prod_{i=1}^{s} \left[ \prod_{j=1}^{r(s)} \frac{\partial}{\partial x^{\mu_{i,j}}} \right] \Psi^{(s)}(i) :$$

where $\Psi^{(s)}$ is one of the fields $\Phi^{(i)}$ of the field theory, or possibly one of the $\Phi^{(i)*}$ if $\Phi^{(i)}$ is non-SCC. Note that (1.54) is quite similar in form to the original interaction Lagrangian (1.18). There are some important differences. We have not limited the number of derivatives of a field which can occur in (1.54); indeed, to renormalize some field theories, the additional terms in the Lagrangian must contain field derivatives of arbitrarily high order. Secondly, the coefficient M of (1.54) contains a factor $g^m$ with $m \geq 2$, whereas our original interaction Lagrangian was first-order in g. Finally, the co-efficients M depend on the regularizing parameters r and $\epsilon$, and will actually, diverge when $r \to 0$. It is for this reason that they are called "infinite counter terms."

In spite of these differences, we may form the series (1.32) with the new Lagrangian just as we did with the old one. If the additions (1.54) are chosen correctly, we may then separate the terms of the new series into sets, one set for each term $\mathcal{T}_{\epsilon,r}(G)$ of the original series, in such a way that the sum of the amplitudes over the set is the renormalized amplitude $\mathcal{R}_{\epsilon,r}(G)$. In a formal sense, then, after the limits $r, \epsilon \to 0$, the new perturbation series is precisely the original one with each Feynman amplitude renormalized.

(ii). Notation.

In this subsection we restate the results of Section 2(C) on the perturbation expansions of covariant truncated TOVEV's, using a more elaborate notation which enables us to describe the graphs involved more explicitly. We begin with the two basic elements of the theory—the set of fields, and the interaction Lagrangian.

The basic fields of the theory are $\Phi^{(1)}, \ldots, \Phi^{(I)}$; suppose that there are $I'$ non-SCC fields among them. Let $\Gamma$ be a set of $I + I'$ elements which indexes the fields and the adjoints of the non-SCC ones; it is clear what we mean by saying that a certain field $\Psi$ is "of type $\gamma$" for $\gamma \in \Gamma$. (We never deal with the adjoints of SCC fields.) Let $\Gamma^R$ be the set of R-tuples $(\gamma_1, \ldots, \gamma_R)$, define $(\gamma_1, \ldots, \gamma_R) \sim (\gamma_1', \ldots, \gamma_R')$ if $\gamma_i' = \gamma_{s(i)}$ for some permutation $s \in S_R$, and let $\Gamma_0^R$ be the set of equivalence classes. (This is a relevant concept because the set of graphs occurring in (1.32) depends only on the equivalence class in $\Gamma_0^R$ associated with the fields $\Psi^{(1)}, \ldots, \Psi^{(R)}$.) Let us give $\Gamma$ some arbitrary ordering; then $\Gamma_0^R$ may also be defined as $\{(\gamma_1, \ldots, \gamma_R) \mid \gamma_1 \leq \gamma_2 \leq \cdots \leq \gamma_R\}$. We will use these two definitions of $\Gamma_0^R$ interchangeably.

We will here let $\Theta$ be the set which indexes the terms in the interaction Lagrangian, that is,

$$\mathcal{L}_I(x) = \sum_{\theta \in \Theta} \mathcal{L}^{(\theta)}(x) = g \sum_{\theta \in \Theta} \sum_{\alpha_1, \ldots, \alpha_{s(\theta)}} M^{(\theta)}_{\alpha_1, \ldots, \alpha_{s(\theta)}} : \sum_{i=1}^{s(\theta)} \Psi^{(\theta, i)}_{\alpha_i}(x) : \tag{1.55}$$

and will sometimes refer to this as the $\Theta$-theory, to distinguish it from other theories with the same fields but different interactions. (Note that (1.55) is just (1.18) with $\Theta$ replacing $\{1, \ldots, Q\}$.) We suppose that $\Psi^{(\theta, i)}$ is of type $\gamma_{\theta, i}$ and that $\gamma_{\theta, 1} \leq \cdots \leq \gamma_{\theta, s(\theta)}$.

Now let us study the covariant truncated TOVEV

$$(\Omega, P^*[\Psi^{(1)}_{\alpha_1}(x_1) \cdots \Psi^{(R)}_{\alpha_R}(x_R)]\Omega)^T \tag{1.56}$$

in the $\Theta$ theory. Suppose that the field $\Psi^{(j)}$ is of type $\gamma_j$, with $\gamma_1 \leq \gamma_2 \leq \cdots \leq \gamma_R$

(this involves no loss of generality), and set $\delta = (\gamma_1, \dots, \gamma_R) \in \Gamma_0^R$. The set of graphs which occurs in the expansion of (1.56) depends only on $\delta$ and $\Theta$. As discussed in Section 2(C. iii), each graph of order m has R + m vertices; the vertices $V_1, \dots, V_R$ being associated with the fields $\Psi^{(1)}, \dots, \Psi^{(R)}$ respectively and the vertices $V_{R+1}, \dots, V_{R+m}$ being associated with terms in the interaction Lagrangian. Such a graph may be described by a pair $(q, \mathcal{L})$ defined as follows:

(a) q is any map $q: \{R+1, \dots, R+m\} \to \Theta$. This map specifies the term in the interaction Lagrangian (1.55), namely $\mathcal{L}^{(q(j))}$, which acts at the vertex $V_j$.

(b) Given q, we define sets $\mathcal{V}_j^q$, for $j = 1, \dots, R+m$, by

$$(1.56) \qquad \mathcal{V}_i^q = \begin{cases} \{j\}, & \text{if } j \le R, \\ \{(j,i) \mid i = 1, \dots, s(q(j))\}, & \text{if } j > R, \end{cases}$$

and set $\mathcal{V}^q = \cup_{j=1}^{R+m} \mathcal{V}_j^q$. The set $\mathcal{V}_j^q$ indexes the fields associated with the vertex $V_j$ of the graph, specifically $\Psi^{(j)}$ for $j \le R$ and $\Psi^{[q(j),1]} \dots \Psi^{[q(j),s(q(j))]}$ for $j > R$. Thus $\mathcal{V}^q$ indexes the fields in the product

$$(1.57) \qquad \prod_{j=1}^{R} \Psi^{(j)}(x_j) \prod_{j=R+1}^{R+m} \mathcal{L}^{(q(j))}(x_j) ,$$

which must be completely contracted to produce the graph. We also define functions $\rho, \pi$, and $\gamma$ on $\mathcal{V}^q$ by

$$\rho(v) = \begin{cases} j, & \text{if } v = j, \\ j, & \text{if } v = (j,i); \end{cases}$$

$$\pi(v) = \begin{cases} j, & \text{if } v = j, \\ (q(j), i), & \text{if } v = (j,i); \end{cases}$$

$$\delta(v) = \text{index (in } \Gamma) \text{ of } \Psi^{(\pi(v))} .$$

Thus the field in (1.57) indexed by v is $\Psi^{(\pi(v))}(x_{\rho(v)})$, and is of type $\gamma(v)$.

(c). The second element of the pair $(q, \mathcal{L})$ is the set of lines of the graph. Such a line arises from the contraction of two fields in (1.57); we therefore regard $\mathcal{L}$ as a subset of $\mathcal{V}^q \times \mathcal{V}^q$. It must satisfy four conditions:

(1). If $(v_1, v_2) \in \mathcal{L}$, then $\rho(v_1) \ne \rho(v_2)$ (this is because the Lagrangian is Wick ordered).

(2). Each $v \in \mathcal{V}^q$ appears exactly once in some pair $(v_1, v_2) \in \mathcal{L}$.

(3). If $(v_1, v_2) \in \mathcal{L}$, the fields of type $\gamma(v_1)$ and $\gamma(v_2)$ have non-zero contraction.

(4). The resulting Feynman graph (described immediately below) is connected.

This leads to

*Definition 1.7:* To each pair $(q, \mathcal{L})$ we assign a Feynman graph. The vertices are $V_1, \ldots, V_{R+m}$, and for each $\ell = (v_1, v_2) \in \mathcal{L}$ there is a line with $i_\ell = \rho(v_1)$, $f_\ell = \rho(v_2)$, and with the propagator

$$\Delta^{(\ell)}_{\alpha_{v_2}\alpha_{v_1}}(x_{\rho(v_2)} - x_{\rho(v_1)}) = \overline{\psi^{(\pi(v_1))}_{\alpha_{v_1}}(x_{\rho(v_1)})\, \psi^{(\pi(v_2))}_{\alpha_{v_2}}(x_{\rho(v_2)})}\,. \tag{1.58}$$

Then the set of all $(q, \mathcal{L})$ satisfying (a), (b) and (c) is denoted $\mathcal{G}(\Theta, \delta, m)$ (but see Remark 1.8); under this correspondence $\mathcal{G}(\Theta, \delta, m)$ *consists of all graphs of order* $m$ *which contribute to* (1.56) *in the* $\Theta$ *theory.*

*Remark 1.8:* (a). We have defined $\mathcal{L}$ as a set of *ordered* pairs; thus, in the graph of Definition 1.7 corresponding to $(q, \mathcal{L})$, each line has a natural orientation. However, Definition 1.2 involves sets of *contractions* of pairs of fields, which does not involve the concept of order. Therefore, *in defining* $\mathcal{G}(\Theta, \delta, m)$, *we do not distinguish between pairs* $(q, \mathcal{L})$ *and* $(q', \mathcal{L}')$ *which differ only in orientation of the lines* (i.e., for which $q = q'$ and, for each $(v_1, v_2) \in \mathcal{L}$, either $(v_1, v_2)$ or $(v_2, v_1) \in \mathcal{L}'$), since these are really the same graph.

(b). On the other hand, it is convenient to have a definite orientation of the lines of a graph. For non-SCC fields one may orient in the direction of flow of charges, but there is no canonical orientation for the propagators of SCC fields. We will therefore assume that we have chosen, once and for all, a fixed (but arbitrary) orientation for the lines in each graph of $\mathcal{G}(\Theta, \delta, m)$; thus we may again describe $\mathcal{G}(\Theta, \delta, m)$ as a set of pairs $(q, \mathcal{L})$, with $\mathcal{L}$ a set of ordered pairs. This choice of orientation does not affect the amplitude defined below.

Having discussed the graphs occurring in the expansion of (1.56), we now turn to their amplitudes and to the series as a whole.

*Definition 1.9:* Take $G = (q, \mathcal{L}) \in \mathcal{G}(\Theta, \delta, m)$. The *regularized amplitude* $A_{\epsilon,r}(G)$ is given by

$$\begin{aligned} A_{\epsilon,r}(G)(x_1, \ldots, x_R)_{\alpha_1,\ldots,\alpha_R} &= \frac{(-ig)^m}{m!} \sum_{\{\alpha_v \mid \rho(v) > R\}} \prod_{j=R+1}^{R+m} M^{(q(j))}_{\alpha_{j,1}\cdots\alpha_{j,s(q(j))}} \\ &\int \cdots \int dx_{R+1} \cdots dx_{R+m}\, \sigma \prod_{\ell \in \mathcal{L}} \left[\Delta^{(\ell)}_{\epsilon,r}\right]_{\alpha_{v_2}\alpha_{v_1}} (x_{\rho(v_2)} - x_{\rho(v_1)}) \end{aligned} \tag{1.59}$$

Here $\Delta^{(\ell)}_{\epsilon,r}$ is the propagator (1.58) regularized as in (1.35) [we take $\ell = (v_1, v_2)$]. $\sigma$ is the sign of the permutation of fermion factors in the reordering

$$\prod_{j=1}^{R} \psi^{(j)} \prod_{j=R+1}^{R+m} \prod_{i=1}^{s(q(j))} \psi^{(q(j),i)} \to \prod_{\mathcal{L}} \psi^{(\pi(v_1))}\psi^{(\pi(v_2))}$$

The *regularized perturbation series in the* $\Theta$ *theory* for (1.56) is then

$$(\Omega, P^*[\Psi^{(1)}_{\alpha_1}(x_1) \cdots \Psi^{(R)}_{\alpha_R}(x_R)]\,\Omega)^T_{\Theta,\epsilon,r} = \sum_{n=0}^{\infty} \sum_{G \in \mathcal{G}(\Theta,\delta,m)} A_{\epsilon,r}(G)(x_1, \ldots, x_R)_{\alpha_1,\ldots,\alpha_R}\,. \tag{1.60}$$

The *regularized renormalized series*,

$$\mathcal{R}(\Omega, P^*[\Pi \Psi^j_{a_j}(x_i)]\Omega)^T_{\Theta,\epsilon,r} , \tag{1.61}$$

is defined similarly, but with all amplitudes renormalized according to Definition 1.5.

*Remark 1.10:* (a). The series (1.60) and (1.61) are well defined term by term. We ignore questions of convergence, of course, but what follows could be made quite rigorous using formal power series in g.

(b). Our results of Section 2(c) are now precisely that the covariant truncated TOVEV (1.56) is the limit of (1.60) as $\epsilon$ and r vanish. Of course, this limit does not exist in general.

(c). Our program of (i) may now be restated: we wish to find a new interaction Lagrangian

$$\mathcal{L}_I'(x) = \sum_{\theta \,\epsilon\, \Theta'} \mathcal{L}^{(\theta)}_{\epsilon,r}(x)$$

with $\Theta \subset \Theta'$, so that

$$\mathcal{R}(\Omega, P^*[\Pi \Psi^{(j)}_{a_j}(x_j)]\Omega)^T_{\Theta,\epsilon,r} = (\Omega, P^*[\Pi \Psi^{(j)}_{a_j}(x_j)]\Omega)^T_{\Theta',\epsilon,r} . \tag{1.62}$$

Note that the right hand side of (1.62) depends on $\epsilon$ and r through the $\epsilon, r$ dependence of $\mathcal{L}^{(\theta)}$ (for $\theta \,\epsilon\, \Theta'-\Theta$) as well as through the regularization of propagators.

We must introduce one more bit of notation. For any graph $G = (q, \mathcal{L}) \,\epsilon\, \mathcal{G}(\Theta, \delta, m)$, it is useful to distinguish the internal structure of the graph (i.e., the generalized vertex $G(V_{R+1}, \ldots, V_{R+m})$) from the lines connecting it to the external vertices. (These latter lines introduce no divergences and do not affect renormalization, since every graph containing one of them is IPR.) We therefore give

*Definition 1.11:* Let $\overline{\mathcal{L}} = \{(v_1, v_2) \,\epsilon\, \mathcal{L} \mid \rho(v_1) > R \text{ and } \rho(v_2) > R\}$ be the set of lines of $G(V_{R+1}, \ldots, V_{R+m})$. Let $\mathcal{U}(G) = \{v \,\epsilon\, \mathcal{U}^q \mid \exists\, (v, v') \text{ or } (v', v) \,\epsilon\, \mathcal{L} \text{ with } \rho(v') \le R\}$, so that $\mathcal{U}(G)$ indexes the fields at internal vertices which are contracted to external fields, and give $\mathcal{U}(G)$ an order in such a way that $v_1 < v_2$ implies $\gamma(v_1) \le \gamma(v_2)$. Let $\beta(G)$ be the collection of indices $a_v$ such that $\rho(v) > R$ and $v \notin \mathcal{U}(G)$. Then define

$$\mathcal{T}_{\epsilon,r}(G)(x_{R+1}, \ldots, x_{R+m})_{\beta(G)} = \prod_{\ell \,\epsilon\, \overline{\mathcal{L}}} [\Delta^{(\ell)}_{\epsilon,r}]_{a_{v_2} a_{v_1}} (x_{\rho(v_2)} - x_{\rho(v_1)}) , \tag{1.63}$$

so that (1.59) becomes

(1.64)

$$A_{\epsilon,r}(G)(x_1, \ldots, x_R)_{a_1,\ldots,a_R} = \frac{(-ig)^m}{m!} \sum_{\{a_v \mid \rho(v) > R\}} \prod_{j=R+1}^{R+m} M^{(q(j))}_{a_{j,1},\ldots,a_{j,s(q(j))}}$$

$$\int \cdots \int dx_{R+1} \cdots dx_{R+m}\, \sigma\, \mathcal{T}_{\epsilon,r}(G)(x_{R+1}, \ldots, x_{R+m})_{\beta(G)} \prod_{(\mathcal{L}-\overline{\mathcal{L}})} [\Delta^{(\ell)}_{\epsilon,r}]_{a_{v_2} a_{v_1}} (x_{\rho(v_2)} - x_{\rho(v_1)}) .$$

(iii). Transformations of graphs.

(a). To keep track of the contributions of various counterterms in the Lagrangian we must study the relationship between amplitudes associated with different graphs in the set $\mathcal{G}(\Theta, \delta, m)$. We do this by studying certain ways of transforming one graph into another. All such transformations will belong to the following general class: Let Y be a group, let $\mathcal{G}$ be a subset of $\mathcal{G}(\Theta, \delta, m)$, and let $S(\mathcal{G})$ be the permutation group on $\mathcal{G}$. We will study homomorphisms $f: Y \to S(\mathcal{G})$. Take $y \in Y$ and $G = (q, \mathcal{L}) \in \mathcal{G}$. The mapping $f(y)$ will always be specified as follows:

(1). First $q' = f(y)(q)$ is defined for each $q$.

(2). Second, a map $f_0(y): \mathcal{V}^q \to \mathcal{V}^{q'}$ is defined.

(3). Finally, since $\mathcal{L} \subset \mathcal{V}^q \times \mathcal{V}^q$, the map $f_0(y)$ defines naturally the image $f(y)(\mathcal{L})$ by

$$f(y)(\mathcal{L}) = [f_0(y) \times f_0(y)](\mathcal{L}) .$$

In all cases considered it will be clear that, for $(q, \mathcal{L}) \in \mathcal{G}$, $[f(y)(q), f(y)(\mathcal{L})]$ is also in $\mathcal{G}$.

(b). We must consider the well-known operation of permuting labels on internal vertices. Let $\mathcal{G} = \mathcal{G}(\Theta, \delta, m)$, and take $Y = S_m$, with $S_m$ considered as a permutation group on $\{R+1, \ldots, R+m\}$. Define $f: S_m \to S(\mathcal{G})$, as above, by

$$f(s)(q) = q \circ s^{-1},$$

$$f_0(s)(v) = \begin{cases} v, & \text{if } \rho(v) \leq R, \\ (s(j), i), & \text{if } v = (j, i) . \end{cases}$$

[To understand this operation, we draw a picture of the graph G. From the vertex $V_k$ $(k > R)$ emerge lines labeled by $\mathcal{V}_k^q = \{(k, 1), (k, 2), \ldots\}$. Then the graph $f(s)G$ is obtained by relabeling this vertex $V_{s(k)}$, and correspondingly relabeling the lines $(s(k), 1), (s(k), 2), \ldots$ . This gives the above action of $f_0$. Note that since this procedure implies $f(q)[s(k)] = q[k]$, we must have $f(q) = q \circ s^{-1}$.] It is clear that, because of the x-integrations in (1.59), we have $A_{\epsilon,r}(G) = A_{\epsilon,r}(f(s)[G])$ for any $s \in S_m$, $G \in \mathcal{G}(\Theta, \delta, m)$. Note also that, because G is connected and has at least one external vertex, we cannot have $f(s)(G) = f(s')(G)$ for $s \neq s'$.

(c). We are also interested in changing the connections between the internal part of a graph and the various external vertices. Suppose $\delta = (\gamma_1, \ldots, \gamma_R) \in \Gamma_0^R$, with $\gamma_1 = \gamma_2 = \cdots = \gamma_{i_1} < \gamma_{i_1+1} = \cdots = \gamma_{i_2} < \gamma_{i_2+1} = \cdots \gamma_R$; we define $T(\delta) \subset S_R$ to consist of all permutations of $\{1, \ldots, R\}$ which leave invariant the sets $\{1, \ldots, i_1\}$, $\{i_1+1, \ldots, i_2\}$ etc. Then take $Y = T(\delta)$ and $\mathcal{G} = \mathcal{G}(\Theta, \delta, m)$, and define $g: Y \to S(\mathcal{G})$ by

$$g(t)(q) = q ,$$

$$g_0(t)(v) = \begin{cases} t(v), & \text{if } \rho(v) \leq R \\ v, & \text{if } \rho(v) > R \end{cases}$$

for any $t \in T(\delta)$. Stated in words, the operation $g(t)$ changes the way the external fields are contracted into the internal structure, without changing internal contractions. Note again that if $g(t)(G) = g(t')(G)$ for some $G \in \mathcal{G}(\Theta, \delta, m)$, we must have $t = t'$. In contrast to the action of

$S_m$, however, the action of T(S) does not, in general, preserve the amplitude $A_{\epsilon,r}(G)$.

(d). We must consider one final operation on graphs. Suppose that a term $\mathcal{L}^{(\theta)}$ in the interaction Lagrangian contains the same field several times, and suppose $\mathcal{L}^{(\theta)}$ acts at the vertex $V_j$ of some graph $G = (q, \mathcal{L})$ [i.e., $q(j) = \theta$]. Then we may change the connections of the lines at $V_j$ without making any other changes in the graph. To make this formal, we proceed as follows. The term $\mathcal{L}^{(\theta)}$ contains the product $\Pi_{i=1}^{s(\theta)} \psi^{(\theta,i)}$ where $\psi^{(\theta,i)}$ is of type $\gamma_{\theta,i}$. Define $U(\theta) = T(\gamma_{\theta,1}, \ldots, \gamma_{\theta,s(\theta)})$ [recall that $T(\delta)$ was defined in (c)]. For any $q_0$: $\{R+1, \ldots, R+m\} \to \Theta$, define

$$U(q_0) = U[q_0(R+1)] \times \cdots \times U[q_0(R+m)] ,$$

$$\mathcal{G}(\Theta, \delta, m, q_0) = \{(q, \mathcal{L}) \in \mathcal{G}(\Theta, \delta, m) \mid q = q_0\} .$$

Then we proceed as in (a), taking $Y = U(q_0)$, $\mathcal{G} = \mathcal{G}(\Theta, \delta, m, q_0)$, and defining $h: Y \to S(\mathcal{G})$ by

$$h(\underline{u})(q) = q \quad (= q_0)$$

$$h_0(\underline{u})(v) = \begin{cases} v, & \text{if } \rho(v) \leq R , \\ (j, u_j(i)), & \text{if } v = (j, i) , \end{cases}$$

for any $\underline{u} = (u_{R+1}, \ldots, u_{R+m}) \in U(q_0)$. Thus the operation $h(\underline{u})$ rearranges the lines at $V_j$ according to $u_j$.

*Remark 1.12:* (a). The operation $h(\underline{u})$ does not necessarily preserve the amplitude $A_{\epsilon,r}(G)$ (i.e., $A_{\epsilon,r}(G) \neq A_{\epsilon,r}[h(\underline{u})(G)]$ in general), but may do so if the Lagrangian has certain symmetries. Thus for $\theta \in \Theta$, Let $U_0(\theta) \subset U(\theta)$ be the subgroup consisting of all u satisfying

$$M^{(\theta)}_{\alpha_1, \ldots, \alpha_{s(\theta)}} = \sigma(u) M^{(\theta)}_{\alpha_{u(1)}, \ldots, \alpha_{u(s(\theta))}} ; \tag{1.65}$$

here $\sigma(u)$ is the sign of the permutation induced by u on the fermion indices among $\alpha_{u(1)}, \ldots, \alpha_{u(s(\theta))}$. Then for any $\underline{u} \in U(q_0)$, with $u_j \in U_0(q(j))$ for all $j = R+1, \ldots, R+m$, one calculates easily

$$A_{\epsilon,r}[h(\underline{u})(G)] = A_{\epsilon,r}(G) .$$

(b). It may be shown that one may, without loss of generality, assume (1.65) for all $\theta \in \Theta$ and $u \in U(\theta)$. We will not need this result, so we omit the simple proof.

(e). We finally must consider the combined action of the groups $S_m$ and $T(\delta)$ (discussed in (b) and (c)). Because the actions of these groups commute, there is a map

$$(f \times g): S_m \times T(\delta) \to S[\mathcal{G}(\Theta, \delta, m)]$$

defined in the obvious way. However, it is important to note that, in contrast to the situation when $S_m$ and $T(\delta)$ act separately, it is possible to have $f(s)g(t)(G) = G$ even if neither s nor t is the identity. For example, in the graph of Figure 1.2, we get this behavior by taking S and t to be the transpositions $3 \longleftrightarrow 4$ and $1 \longleftrightarrow 2$, respectively, if we assume, say, $\mathcal{L} = \{[(3,1),1],[(4,1),2],[(3,2),(4,2)],[(3,3),(4,3)],[(3,4),(4,4)]\}$.

Figure 1.2

Having made these remarks, we give

*Definition 1.13:* For $G \in \mathcal{G}(\Theta, \delta, m)$, define the *diagram*

$$\bar{G} = \{G' \in \mathcal{G}(\Theta,\delta,m) \mid G' = f(s)g(t)G, \text{ for some } (s,t) \in S_m \times T(\delta)\} ,$$

and choose once and for all a single graph $G^* \in \bar{G}$; $G^* = (q^*, \mathcal{L}^*)$. The set of all diagrams $\bar{G} \subset \mathcal{G}(\Theta,\delta,m)$ is a partition of $\mathcal{G}(\Theta,\delta,m)$. (Diagrams will play an important role in what follows.) Define a subgroup $T_0(\bar{G}) \subset S_m \times T(\delta)$ by

$$T_0(\bar{G}) = \{(s,t) \mid f(s)g(t)(G^*) = G^*\} ,$$

and set $d(\bar{G}) = \#[T_0(\bar{G})]$. (If $G^*$ is the graph of Figure 1.2, $T_0(\bar{G}) = \{[e,e],[(1\ 2),(3\ 4)]\}$, where e is the identity and (a b) denotes the transposition of a and b.) We will also need the subgroup of $T(\delta)$. given by

$$T_1(\bar{G}) = \{t \in T(\delta) \mid (s,t) \in T_0(\bar{G}), \text{ some } s \in S_m\} .$$

*Remark 1.14:* (a). Note two trivial consequences of this definition:

(i). For any $G' \in \bar{G}$,

$$\#[\{(s,t) \in S_m \times T(\delta) \mid f(s)g(t)G^* = G'\}] = d(\bar{G}) ;$$

(ii).

$$\#[T_1(\bar{G})] = d(\bar{G}) .$$

(b). Our terminology of "graph" and "diagram" is not standard; no standard terminology exists. Our graphs index *all* terms arising from Wick's theorem; such objects are more often referred to as "labeled graphs." The term "graph" or "diagram" usually refers to graphs with all labels removed (except for labels identifying the type of propagator); that is, to the topological structure underlying one of our graphs. This corresponds to a partition of $\mathcal{G}(\Theta, \delta, m)$ into even fewer subsets than in the partition given by our diagrams; specifically, elements of $\mathcal{G}(\Theta,\delta,m)$ which are related by transformations of the type considered in (b), (c), or (d) are grouped together.

(iv) Counterterms.

In this subsection we define the new interaction Lagrangian $\mathcal{L}_I'$, discussed in Remark 1.10(c). In subsection (v) we will conclude our discussion by verifying the equation (1.62).

In order to see how to define $\mathcal{L}_I'$, we will group the terms in the series (1.60) into sums over the diagrams $\bar{G}$ of Definition 1.13. Thus we have

$$(1.66)\quad \left(\Omega, P^*\left[\prod_1^R \Psi_{a_j}^{(j)}(x_j)\right]\Omega\right)^T_{\Theta,\epsilon,r} = \sum_{m=0}^{\infty} \sum_{\bar{G} \subset \mathcal{G}(\Theta,\delta,m)} B_{\epsilon,r}(\bar{G})(x_1,\ldots,x_R)_{a_1,\ldots,a_R},$$

where for $\bar{G} \subset \mathcal{G}(\Theta,\delta,m)$, we define

$$B_{\epsilon,r}(\bar{G}) = \sum_{G' \epsilon \bar{G}} A_{\epsilon,r}(G').$$

Using Remark 1.14 (a.i) and the invariance of the amplitude $A_{\epsilon,r}(G)$ under action of $S_m$, we obtain

$$B_{\epsilon,r}(\bar{G}) = \frac{1}{d(\bar{G})} \sum_{(s,t)\,\epsilon\, S_m \times T(\delta)} A_{\epsilon,r}[f(s)\,g(t)(G^*)] = \frac{m!}{d(\bar{G})} \sum_{t\,\epsilon\, T(\delta)} A_{\epsilon,r}[g(t)(G^*)]$$

$$(1.67)\quad = \frac{(-ig)^m}{d(\bar{G})} \sum_{\{a_v | \rho(v) > R\}} \prod_{j=R+1}^{R+m} M^{q^*(j)}_{a_{j,1},\ldots,a_{j,s(q^*(j))}} \int\cdots\int dx_{R+1}\cdots dx_{R+m}$$

$$\sigma'(G^*)\,\mathcal{T}_{\epsilon,r}(G^*)(x_{R+1},\ldots,x_{R+m})_{\beta(G^*)}\left(\omega, P^*\left[\prod_{j=1}^R \psi_{a_j}^{(j)}(x_j) : \prod_{\mathfrak{U}(G)} \psi_{a_v}^{(\pi(v))}(x_{\rho(v)}):\right]\omega\right)_{\epsilon,r}$$

where $\sigma'(G^*)$ is the fermion sign factor for the reordering

$$\prod_{j=R+1}^{R+m} \prod_{i=1}^{s(q^*(j))} \psi^{(q^*(j),i)} \longrightarrow \prod_{(v_1,v_2)\,\epsilon\, \bar{\mathcal{L}}^*} \psi^{(\pi(v_1))}\psi^{(\pi(v_2))} \prod_{v\,\epsilon\, \mathfrak{U}(G)} \psi^{(\pi(v))}$$

(see Definition 1.11). The $\epsilon, r$ subscripts on the VEV in (1.66) imply that all propagators are to be regularized.

Now in the notation of (c) $\mathcal{T}_{\epsilon,r}(G^*)$ would be written $\mathcal{T}_{\epsilon,r}(V_{R+1},\ldots,V_{R+m})$. Thus the renormalized series (1.61) is obtained from (1.66) as follows: in each summand $B_{\epsilon,r}(G)$, we replace the quantity $\mathcal{T}_{\epsilon,r}(G^*)$ by the quantity $\mathcal{R}_{\epsilon,r}(V_{R+1},\ldots,V_{R+m})$, as defined in Definition 1.5. We pointed out in (c) that, when $m \geq 2$, $\mathcal{R}_{\epsilon,r}$ is the sum of various terms, one of which is $\mathcal{T}_{\epsilon,r}(V_{R+1},\ldots,V_{4+m})$, and one of which is $\mathcal{X}_{\epsilon,r}(V_{R+1},\ldots,V_{R+m})$. We write $\mathcal{X}_{\epsilon,r}(V_{R+1},\ldots,V_{R+m}) = \mathcal{X}_{\epsilon,r}(G^*)$; $\mathcal{X}_{\epsilon,r}(G^*)$ depends on the same indices $\beta(G^*)$ as does $\mathcal{T}_{\epsilon,r}(G^*)$.

Suppose that we consider an interaction Lagrangian

$$(1.68)\quad \mathcal{L}^{(\bar{G})}_{\epsilon,r}(x) = g\,\frac{(-ig)^{m-1}}{d(\bar{G})} \sum_{\{a_v | \rho(v) > R\}} \prod_{j=R+1}^{R+m} M^{q^*(j)}_{a_{j,1}\cdots a_{j,s(q^*(j))}}$$

$$\times \left\{\int\cdots\int dx_{R+2}\cdots dx_{R+m}\,\sigma'(G^*)\,\mathcal{X}_{\epsilon,r}(G^*)_{\beta(G^*)} : \prod_{v\,\epsilon\,\mathfrak{U}(G^*)} \Psi_{a_v}^{(\pi(v))}(x_{\rho(v)}) : \right\}_{x_{R+1}=x}$$

The lowest order contribution of (1.68) to the TOVEV (1.56) is

$$(1.69) \qquad (-i)\left(\omega, P^*\left[\prod_{j=1}^{R} \psi_{a_j}^{(j)}(x_j)\cdot \mathcal{L}_{\epsilon,r}^{(\bar{G})}(x)\right]\omega\right),$$

and when the propagators in (1.69) are regularized, this is equal to (1.67) with $\mathcal{T}_{\epsilon,r}(G^*)$ replaced by $\mathcal{X}_{\epsilon,r}(G^*)$. Thus if we add (1.68) to our original Lagrangian we will implement at least part of the renormalization. This motivates

*Definition 1.15:* Let $\Theta'$ be the set of all diagrams $\bar{G}$ of the $\Theta$ theory,

$$\bar{G} \subset \bigcup_{m=1}^{\infty} \bigcup_{R=1}^{\infty} \bigcup_{\delta \in \Gamma_0^R} \mathcal{G}(\Theta, \delta, m) ,$$

For $\bar{G} \in \Theta'$, define $\mathcal{L}_{\epsilon,r}^{(\bar{G})}$ by (1.68). Then the *new interaction Lagrangian is*

$$\mathcal{L}'_{I_{\epsilon,r}}(x) = \sum_{\Theta'} \mathcal{L}_{\epsilon,r}^{(\bar{G})}(x) .$$

*Remark 1.16:* (a). It is easy to see that for diagrams $\bar{G} \in \Theta'$, with $\bar{G} \subset \mathcal{G}(\Theta, \delta, 1)$, we have

$$\mathcal{L}_{\epsilon,r}^{(\bar{G})}(x) = \mathcal{L}^{(q^*(R+1))}(x)$$

(recall $G^* = (q^*, \mathcal{L}^*)$). Thus $\Theta \subset \Theta'$ in a natural way.

(b). We have already remarked that the lowest order contribution of (1.68) (with $m \geq 2$) provides one term in the renormalization of (1.67). In (v) below we will show that the $\Theta'$ defined in Definition 1.15 actually gives all renormalization, i.e., satisfies (1.62).

(c). It may have been noticed that (1.68) is not of the promised form (1.54). Recall, however, that $\mathcal{X}_{\epsilon,r}(G^*)$ is actually a distribution of the form

$$(1.70) \quad \mathcal{X}_{\epsilon,r}(G^*)(x_{R+1}, \ldots, x_{R+m}) = \left[Z_{\epsilon,r}(G^*)\left(\frac{\partial}{\partial x_j}\right)\right]\delta(x_{R+1}-x_{R+2})\cdots\delta(x_{R+m-1}-x_{R+m}) ,$$

where $Z_{\epsilon,r}(G^*)$ is a polynomial in the derivatives $(\partial/\partial x_i)$. (We have suppressed the dependence on the indices $\beta(G^*)$.) Thus, aside from coefficients, the contribution of (1.68) to any amplitude has the form

$$(1.71) \quad \int \cdots \int dx_{R+1} \cdots dx_{R+m}\left\{\left[Z_{\epsilon,r}(G^*)\left(\frac{\partial}{\partial x_i}\right)\right]\delta(x_{R+1}-x_R)\cdots \prod_{v \in \mathcal{U}(G^*)} \psi_{a_v}^{(\pi(v))}(x_{\rho(v)})\,\psi^{(?)} ,\right.$$

where $\psi^{(?)}$ is some field not evaluated at any of $x_{R+1}, \ldots, x_{R+m}$. We may thus integrate by parts in (1.71), transferring derivatives from the $\delta$-functions to the fields $\psi^{\pi(v)}$, and then do the integrals over $x_{R+2}, \ldots, x_{R+m}$ explicitly. This means that (1.68) is equivalent to

$$\mathcal{L}^{(\bar{G})}_{\epsilon,r}(x) = g\,\frac{(-ig)^{m-1}}{d(\bar{G})} \sum_{\{\alpha_v|\rho(v)>R\}} \prod_{j=R+1}^{R+m} M^{(q^*(j))}_{\alpha_{j,1}\cdots\alpha_{j,s(q^*(j))}}$$

$$\left\{ \left[ Z_{\epsilon,r}(G^*)_{\beta(G^*)}\left(-\frac{\partial}{\partial x_j}\right) \right] : \prod_{v\in\mathcal{U}(G^*)} \Psi^{(\pi(v))}_{\alpha_v}(x_{\rho(v)}) : \right\}_{x_{R+1}=\cdots=x_{R+m}=x}$$

which has the desired form (1.54).

(d). Actually, Definition 1.15 formally adds more terms to the Lagrangian than are necessary. This is because $\mathcal{L}^{(\bar{G})} = 0$ unless

(i) $G^*(V_{R+1},\ldots,V_{R+m})$ is IPI,

(ii) $\mu[G^*(V_{R+1},\ldots,V_{R+m})] \geq 0$.

We have defined $\Theta'$ this way only to avoid discussing special cases later.

(v). Renormalization of the perturbation series.

*Theorem 1.17:* Suppose that we have a field theory (the $\Theta$ theory) and let $\Theta'$ be given by Definition 1.15. Then for any fields $\Psi^{(1)},\ldots,\Psi^{(R)}$, the regularized series (1.60) in the $\Theta'$ theory is equal to the regularized renormalized series in the $\Theta$-theory:

$$\text{(1.72)}\qquad \mathcal{R}\left(\Omega, P^*\left[\prod_1^R \Psi^{(j)}_{\alpha_j}(x_j)\right]\Omega\right)^T_{\Theta,\epsilon,r} = \left(\Omega, P^*\left[\prod_1^R \Psi^{(j)}_{\alpha_j}(x_j)\right]\Omega\right)^T_{\Theta',\epsilon,r}.$$

(See the discussion in Remark 1.10 (c) of the $\epsilon$ and $r$ dependence of the right hand side of (1.72).)

*Proof:* There is no one-one correspondence between the terms in the series on the two sides of (1.72). Rather, we will construct a correspondence between certain sets of terms, and show that the sum of the terms in such a set on one side is equal to the sum of the terms in the corresponding set on the other side. We divide the proof into several sections.

(a). Let us define $\mathcal{H}(\Theta,\delta,m)$ to be the set of all pairs $(G,P)$, with $G$ a graph in $\mathcal{G}(\Theta,\delta,m)$ and $P$ a partition of $\{R+1,\ldots,R+m\}$ into $k(P)$ sets $U^P_k = \{j^P_{k,1},\ldots,j^P_{k,m_k}\}$, with $k = R+1,\ldots,R+k(P)$. We will adopt the numbering convention that $j^P_{R+1,1} < j^P_{R+2,1} < \cdots < j^P_{R+k(P),1}$, and $j^P_{k,1} < \cdots < j^P_{k,m_k}$. For $(G,P)\in\mathcal{H}(\Theta,\delta,m)$, we define an amplitude $A_{\epsilon,r}(G,P)$ to be given by (1.64) with $\mathcal{I}_{\epsilon,r}(G)$ replaced by

$$\prod_{k=R+1}^{R+k(p)} \mathcal{X}_{\epsilon,r}(V_{j^P_{k,1}}\cdots V_{j^P_{k,r(k)}}) \prod_{\text{conn}} \Delta^{(\ell)}_{\epsilon,r},$$

where $\prod_{\text{conn}}$ is the product over all lines of $\bar{\mathcal{L}}$ which join vertices lying in different $U_k$.

Then from Definitions 1.5 and 1.9, the left hand side of (1.72) is

$$\text{(1.73)}\qquad \mathcal{R}\left(\Omega, P^*\left[\prod_1^R \Psi^{(j)}_{\alpha_j}(x_j)\right]\Omega\right)^T_{\Theta,\epsilon,r} = \sum_{m=0}^{\infty} \sum_{(G,P)\in\mathcal{H}(\Theta,\delta,m)} A_{\epsilon,r}(G,P)(x_1,\ldots,x_R)_{\alpha_1\cdots\alpha_R}.$$

In (iii, b) we defined a map $f\colon S_m \to S[\mathcal{G}(\Theta,\delta,m)]$. In an obvious way we extend this to a new map $f'\colon S_m \to S[\mathcal{H}(\Theta,\delta,m)]$, defined by $f'(s)(G,P) = [f(s)(G), f'(s)(P)]$ where, if $P = \{U_k^P\}_{k=1}^{k(P)}$, $f'(s)(P) = \{s(U_k^P)\}_{k=1}^{k(P)}$. [To understand the mapping $f'$ intuitively, consider Figure 1.3. As indicated, $P$ is the partition with $U_1^P = \{4,5\}$, $U_2^P = \{6\}$. Let $s$ be the cyclic permutation $(4\ 5\ 6)$; the partitioned graph

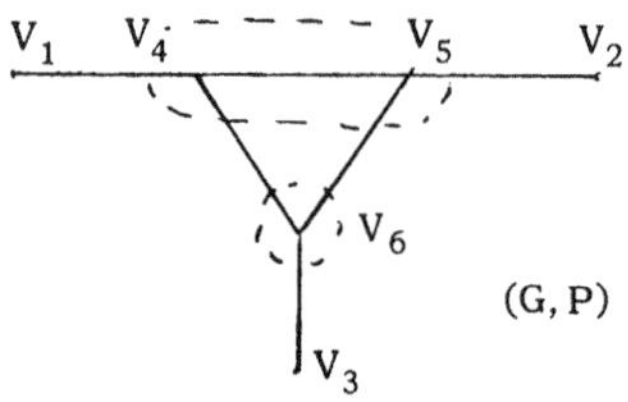


Figure 1.3

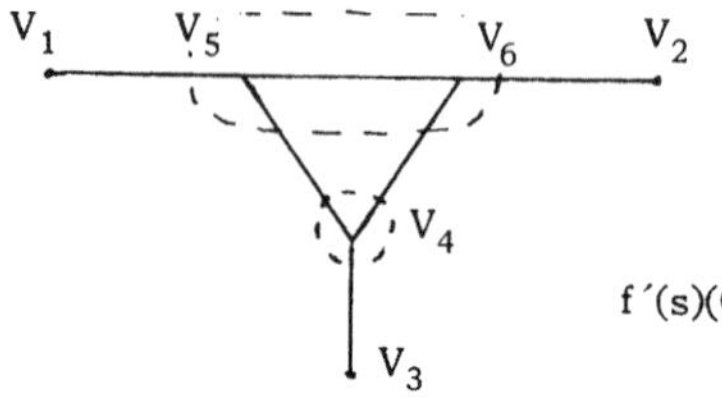


Figure 1.4

$f'(s)(G,P)$ is shown in Figure 1.4.] As in (iii,b), we have $A_{\epsilon,r}[f'(s)(G,P)] = A_{\epsilon,r}[G,P]$. Thus, if we set $\widetilde{(G,P)} = \{(G',P')\,|\,(G',P') = f'(s)(G,P),\ \text{some } s \in S_m\}$, equation (1.73) becomes

(1.74)

$$\mathcal{R}\Big(\Omega, P^*\Big[\prod_1^R \psi_{a_j}^{(j)}(x_j)\Big]\Omega\Big)^T_{\Theta,\epsilon,r} = \sum_{m=0}^{\infty} m! \sum_{\widetilde{(G,P)} \subset \mathcal{H}(\Theta,\delta,m)} A_{\epsilon,r}(G',P')(x_1,\ldots,x_R)_{a_1,\ldots,a_R}.$$

Here $(G',P')$ is any element of $\widetilde{(G,P)}$, and we have used the equation $\#\widetilde{(G,P)} = m!$

(b). Let $\bar{G}$ be a diagram in $\Theta'$, with $\bar{G} \subset \mathcal{G}(\Theta,\delta',m')$, $\delta' \in \Gamma_0^{R'}$. Recall three concepts previously defined:

(1). In Definition 1.13 we chose a distinguished graph $G^* = (q^*, \mathcal{L}^*) \in \bar{G}$.

(2). In (iii,c) we defined $T(\delta')$, a group of permutations on $\{1,\ldots,R'\}$, and described its action on $G^*$.

(3). In (iii,d) we defined a group $U(\bar{G})$; it is a permutation group on $\mathcal{U}(G^*)$ [because the fields in $\mathcal{L}_{\epsilon,r}^{(\bar{G})}$ are indexed by $\mathcal{U}(G^*)$].

We define a map $\tau(G^*)\colon \{1,\ldots,R'\} \to \mathcal{U}(G^*)$ by

$$\tau(G^*)(j) = v \ \text{ iff } \ (v,j) \text{ or } (j,v) \in \mathcal{L}^*,$$

and a map $\eta\colon T(\delta') \to U(\bar{G})$ by

$$\eta(t) = \tau(G^*)\, t\, \tau(G^*)^{-1} .$$

$\eta$ is easily seen to be an isomorphism.

We pause to remark on the intuitive meaning of $\eta$. Consider the graph obtained from $G^*$ by shrinking the generalized vertex $G^*(V_{R'+1},\ldots,V_{R'+m'})$ to a point. The lines of this graph are described by pairs $(j,v)$ [or $(v,j)$], with $j \in \{1,\ldots,R'\}$, $v \in \mathcal{U}(G^*)$. Suppose, for simplicity, that all have the form $(j,v)$. Then this set of lines is preserved under the mapping $t \times \eta(t)$:

$$[t \times \eta(t)] : [\{1,\ldots,R'\} \times \mathcal{U}(G^*)] \to [\{1,\ldots,R'\} \times \mathcal{U}(G^*)] .$$

Finally, recall the definitions of $T_1(\bar{G})$ [Dfn. 1.13] and $U_0(\bar{G})$ (iii,d). For any $t \in T_1(\bar{G})$ there is an $s \in S_{m'}$ such that $f(s)\,g(t)(G^*) = G^*$. We draw two conclusions:

(1) For any $v \in \mathfrak{U}(G^*)$, we have

$$f_0(s)(v) = \eta(t)(v) \ . \tag{1.75}$$

(2) Since the action of $t$ on $G^*$ is equivalent to the action of $s^{-1}$, and since $f(s^{-1})$ preserves amplitudes, we have $\eta(t) \in U_0(\bar{G})$, i.e.,

$$\eta[T_1(\bar{G})] \subset U_0(\bar{G}) \ . \tag{1.76}$$

(c). Let $H$ be any element of $\mathcal{G}(\Theta', \delta, m)$, with $H = (q_1, \mathcal{L}_1)$, and set $U_1(q_1) = \Pi_{j=R+1}^{R+m} \eta[T_1(q_1(j))]$. Define a set of graphs

$$\tilde{H} = \{H' \in \mathcal{G}(\Theta', \delta, m) \mid H' = f(s)h(\underline{u})H, \text{ some } s \in S_m \text{ and } \underline{u} \in U_1(q_1)\} \ .$$

In a moment we will prove the

*Claim:* $\#(\tilde{H}) = m!\ \Pi_{j=R+1}^{R+m} d[q_1(j)]$.

Accepting this, we note that from (1.76) and the fact that the action of $S_m$ always preserves amplitudes, we have

$$A_{\epsilon,r}(H') = A_{\epsilon,r}(H)$$

for any $H' \in \tilde{H}$. The claim then implies that the right hand side of (1.72) is given by

$$\left(\Omega, P^*\left[\prod_1^R \psi^{(j)}_{a_j}(x_j)\right]\Omega\right)^T_{\Theta',\epsilon,r} = \sum_{m=0}^{\infty} \sum_{\tilde{H} \subset \mathcal{G}(\Theta',\delta,m)} m! \prod_{j=R+1}^{R+m} d[q_1(j)]\, A_{\epsilon,r}(H) \ . \tag{1.77}$$

We now prove the claim. Since $\#(S_m) = m!$ and $\#[\eta(T_1[q_1(j)])] = d[q_1(j)]$, we must show that, whenever $f(s)h(\underline{u})H = H$, $s$ and $\underline{u}$ are the identity elements of $S_m$ and $U_1(q_1)$, respectively. Let us call two vertices of a graph $(q, \mathcal{L})$ *adjacent* if some line joins them, and call a vertex $V_j$ *fixed* under a transformation $f(y)$ of the type discussed in (iii,a) if $f(y)(q)(j) = q(j)$, and $f_0(y) \mid \mathcal{O}_j^q$ is the identity on $\mathcal{O}_j^q$. Now note that for any $\bar{G}$ as in (b), $t' \in T_1(\bar{G})$ implies that $t'(i) \neq i$ for any $i \in \{1,\ldots,R'\}$. This is because there is an $s' \in S_{m'}$ with $f(s')g(t')(G^*) = (G^*)$; if any vertex is fixed under $f(s')g(t')$, so are its adjacent ones. But the graph is connected, so either all or no vertices are fixed under $f(s')g(t')$. Now suppose $f(s)h(\underline{u})H = H$ for $s \in S_m$, $\underline{u} \in U_1(q_1)$. By the above property of $T_1(\bar{G})$, we again have that if any vertex of $H$ is fixed under $f(s)h(\underline{u})$, so are its adjacent vertices. But here the external vertices are fixed, so $H$ connected implies that all vertices are fixed. This means that $s$ and $\underline{u}$ are the identities.

(d). Now define

$$\mathcal{I} = \bigcup_{m=0}^{\infty} \{\widetilde{(G,P)} \subset \mathcal{H}(\Theta,\delta,m)\} ,$$

$$\mathcal{J} = \bigcup_{m=0}^{\infty} \{\bar{\mathcal{H}} \subset \mathcal{G}(\Theta',\delta,m)\} .$$

These are the families of sets mentioned at the beginning of the proof. We will now construct a 1-1, onto correspondence $\phi : \mathcal{I} \to \mathcal{J}$ such that, if $\widetilde{(G,P)} \in \mathcal{H}(\Theta,\delta,m)$, $\phi\widetilde{(G,P)} \in \mathcal{G}(\Theta',\delta,k(P))$, and such that for any $(G',P') \in \widetilde{(G,P)}$ and $H' \in \phi\widetilde{(G,P)}$, we have

$$m!\; A_{\epsilon,r}(G',P') = k(P)! \prod_{j=R+1}^{R+m} d(\bar{G}_j) A_{\epsilon,r}(H') . \tag{1.78}$$

Then (1.74), (1.77), and (1.78) imply the theorem.

We begin the construction of $\phi$ by defining a map

$$\psi : \bigcup_{m=0}^{\infty} \mathcal{H}(\Theta,\delta,m) \to \bigcup_{m=0}^{\infty} \mathcal{G}(\Theta',\delta,m) .$$

Intuitively speaking, $\psi$ is the map which, applied to (G, P), replaces each generalized vertex $U_k^P$ by the corresponding term $\mathcal{L}^{(\theta)}$, with $\theta \in \Theta'$. We then show that $\widetilde{\psi(G,P)}$ depends only on $\widetilde{(G,P)}$, and thus define $\phi$ by $\phi\widetilde{(G,P)} = \widetilde{\psi(G,P)}$. (1.78) follows easily once $\phi$ has been properly defined.

Now suppose $(G,P) \in \mathcal{H}(\Theta,\delta,m)$, with $G = (q,\mathcal{L})$ and $P = \{U_k^P\}_{k=R+1}^{R+k(P)}$. Let $\mathcal{V}^{(q,k)}$, for $k = R+1,\dots,R+k(P)$, be the subset of $\mathcal{V}^q$ consisting of all v connected to or in the generalized vertex $U_k^P$:

$$\mathcal{V}^{(q,k)} = \left[\bigcup_{j \in U_k^P} \mathcal{V}_j^q\right] \cup \{v \in \mathcal{V}^q \mid (v,v') \text{ or } (v',v) \in \mathcal{L}, \text{ for some } v' \text{ with } \rho(v') \in U_k^P\} .$$

We may find $\bar{G}_k \in \Theta'$, with corresponding distinguished graph $G_k^* = (q_k^*, \mathcal{L}_k^*) \in \mathcal{G}(\Theta',\delta_k,m_k)$, and $\delta_k \in \Gamma_0^{R_k}$, and maps

$$e_k : U_k^P \to \{R_k+1,\dots,R_k+m_k\} ,$$

$$e_k^0 : \mathcal{V}^{(q,k)} \to \mathcal{V}^{q_k^*} ,$$

such that

i) $q(j) = q_k(e_k(j))$, for any $j \in U_k^P$ ;

ii) $e_k^0(j,i) = (e_k(j),i)$, for any $j \in U_k^P$, $i = 1,\dots,s(q(j))$;

iii) $[e_k^0 \times e_k^0][\mathcal{L} \cap (\mathcal{V}^{(q,k)} \times \mathcal{V}^{(q,k)})] = \mathcal{L}_k^*$ .

That is, the generalized vertex $U_k^P$, with its attached lines, corresponds precisely to $G_k^*$ with some numbering of the vertices. Then we define $\psi(G,P) = (q_1, \mathfrak{L}_1)$ by

$$q_1(k) = \bar{G}_k, \quad k = R+1,\dots,R+k(P),$$

$$\mathfrak{L}_1 = \{(v,v') \in \mathcal{O}^{q_1} \times \mathcal{O}^{q_1} \mid [d^0(v), d^0(v')] \in \mathfrak{L}\}$$

where $d^0\colon \mathcal{O}^{q_1} \to \mathcal{O}^q$ is given by

$$d^0(v) = \begin{cases} v & \text{if } \rho(v) \le R \\ e_k^{0-1}(u) & \text{if } v = (k,u), \quad u \in \mathcal{U}(G_k^*) . \end{cases}$$

We must now verify that

(1). $\widetilde{\psi(G,P)} = \widetilde{\psi(G',P')}$ for any $(G',P') \in \widetilde{(G.P)}$ ;

(2) $\phi$ [defined using (1)] is 1-1 and onto;

(3). equation (1.78) holds.

Statement (3) follows directly from the definitions of the amplitudes. We will prove (1) explicitly, and leave the proof of (2) to the reader.

Thus let $(G',P') = f'(s)(G,P)$, with $s \in S_m$. We assign primes to the quantities $d^0$, $e_k^0$, $\bar{G}_k$, $G_k^*$ when they refer to $(G',P')$. By definition of $P'$, we have $\{U_k^{P'}\} = \{s(U_k^P)\}$, thus we may define $\sigma \in S_{k(P)}$ by

$$U_{\sigma(k)}^{P'} = s(U_k^P) .$$

We clearly have $G_k^* = G_{\sigma(k)}^{*\prime}$. If we define $\alpha_k\colon \mathcal{O}^{q_k^*} \to \mathcal{O}^{q_k^*}$ by

$$\alpha_k = e_{\sigma(k)}^{0\prime} f_0(s) e_k^{0-1} ,$$

then there exist $t_k \in T(\delta_k)$, $s_k \in S_{m_k}$ with $\alpha_k = f_0(s_k) g_0(t_k)$. Since $\alpha_k$ takes $G_k^*$ into itself, we have $t_k \in T_1(\delta_k)$. We claim

$$\psi[f'(s)(G,P)] = f(\sigma) \Pi\, \eta(t_k) [\psi(G,P)] , \tag{1.79}$$

which, if true, proves (1).

Let the left and right hand sides of (1.79) be the graphs $(q_2, \mathfrak{L}_2)$ and $(q_3, \mathfrak{L}_3)$, respectively. We have, for $k = R+1,\dots,R+k(P)$,

$$q_2(k) = \bar{G}_k' = \bar{G}_{\sigma^{-1}(k)} = q_1(\sigma^{-1}(k)) = q_3(k) .$$

Thus, there remains to show only that $\mathfrak{L}_2 = \mathfrak{L}_3$. Take $v_1, v_2 \in \mathcal{O}^{q_2}$ $(= \mathcal{O}^{q_3})$. We will suppose for simplicity that $\rho(v_1) > R$, $\rho(v_2) > R$ (other cases are similar), so that $v_1 = (k_1,u_1)$ $v_2 = (k_2,u_2)$, with $u_i \in \mathcal{U}(G_{k_i}^{*\prime})$. Then

$$\begin{aligned}(v_1,v_2) \in \mathfrak{L}_2 &\iff [d^{0\prime}(v_1), d^{0\prime}(v_2)] \in \mathfrak{L}' \\ &\iff [f_0(s)^{-1}[d^{0\prime}(v_1)], f_0(s)^{-1}[d^{0\prime}(v_2)] \in \mathfrak{L} \\ &\iff (d^{0-1}\{f_0(s)^{-1}[d^{0\prime}(v_1)]\}, d^{0-1}\{f_0(s)^{-1}[d^{0\prime}(v_1)]\}) \in \mathfrak{L}_1 .\end{aligned}$$

But for $i = 1, 2,$

$$d^{0^{-1}}\{f_0(s)^{-1}[d^{0'}(v_i)]\} = [\sigma^{-1}(k_i),\ \alpha^{-1}_{\sigma^{-1}(k_i)}(u_i)]$$

$$= [\sigma^{-1}(k_i),\ f_0(s^{-1}_{\sigma^{-1}(k_i)})(u_i)]$$

$$= [\sigma^{-1}(k_i),\ \eta(t^{-1}_{\sigma^{-1}(k_i)})u_i]$$

by (1.75). But we have precisely

$$\{f_0(\sigma)\,\Pi\, h_0[\eta(t_k)]\}\,[\sigma^{-1}(k_i),\ \eta(t^{-1}_{\sigma^{-1}(k_i)})\,u_i] = (k_i, u_i)\ ,$$

hence $(v_1, v_2) \in \mathfrak{L}_2 \Longleftrightarrow (v_1, v_2) \in \mathfrak{L}_3$. This completes the proof.

E. Finite Renormalization and Physical Interpretation.

One addition to the preceding two sections is necessary. Formulas (1.49)–(1.51) do not, in fact, represent the only possible renormalization of the Feynman amplitude (1.34). Our motivation for (1.51) could be stated as follows: "The coefficients of the Taylor series for $\bar{\mathcal{R}}(v_1', \ldots, v_m')$ up to order $\mu(v_1', \ldots, v_m')$ diverge when $r \to 0$; I will remove this divergence by setting them equal to zero." This is clearly quite arbitrary; the most general requirement would be that these coefficients remain finite. This leads to

*Definition 1.18:* Let G be a Feynman graph. A *finite renormalization* is a map which, for each generalized vertex $U = \{V_1', \ldots, V_m'\}$ of G, gives a distribution $\hat{\mathcal{X}}_\epsilon(U) \in \mathcal{S}'(R^{4m})$ of the form

$$(1.80)\qquad \hat{\mathcal{X}}_\epsilon(U) = \begin{cases} 1 & \text{if } m' = 1, \\ 0 & \text{if } G(V_1', \ldots, V_m') \text{ is IPR}, \\ \delta(\Sigma^m_{i=1} p_i')\, Z_U(p_i') & \text{otherwise.} \end{cases}$$

Here $Z_U$ is a polynomial of degree $\mu(V_1', \ldots, V_m')$ whose coefficients have finite limits when $\epsilon \to 0$. (The possibility of $\epsilon$-dependence of $\hat{\mathcal{X}}_\epsilon(U)$ is included for later convenience, but is not of primary importance. What is important is $\lim_{\epsilon \to 0} \hat{\mathcal{X}}_\epsilon(U)$.)

*Definition 1.19:* Given a Feynman graph G and a finite renormalization $\hat{\mathcal{X}}$ as above, we define a new renormalized amplitude for G by

$$(1.81)\qquad \mathcal{X}'_{\epsilon,r}(V_1', \ldots, V_m') = \begin{cases} 1 & \text{if } m = 1, \\ 0 & \text{if } G(V_1', \ldots, V_m') \text{ is IPR}, \\ -\mathfrak{M}_{\mu(V_1', \ldots, V_m')}\bar{\mathcal{R}}'_{\epsilon,r}(V_1', \ldots, V_m') + \hat{\mathcal{X}}_\epsilon(V_1', \ldots, V_m'); & \text{otherwise.} \end{cases}$$

$$(1.82)\qquad \bar{\mathcal{R}}'_{\epsilon,r}(V_1', \ldots, V_m') = \sum_P \prod_{j=1}^{k(P)} \mathcal{X}'_{\epsilon,r}(V^P_{j,1}, \ldots, V^P_{j,r(j)}) \prod_{\text{conn}} \Delta^{(\ell)}_{\epsilon,r}\ ;$$

(1.83) $$\mathcal{R}'_{\epsilon,r}(V'_1,\ldots,V'_m) = \bar{\mathcal{R}}'_{\epsilon,r}(V'_1,\ldots,V'_m) + \mathcal{X}'_{\epsilon,r}(V'_1,\ldots,V'_m) \ .$$

Here $\Sigma_P$ and $\prod_{\text{conn}}$ defined as in Definition 1.5. Thus in (1.81)–(1.83) we may specify arbitrarily (through the choice of $\hat{\mathcal{X}}$) the first $\mu(V'_1,\ldots,V'_m)$ terms of the Taylor series of the renormalized amplitude for $G(V'_1,\ldots,V'_m)$.

So far we have been dealing with the renormalization of a single graph, but similar modifications may also be made in a field theory. We follow the notation of (D). Thus, if $\bar{G} \in \Theta'$ but $\bar{G} \notin \Theta$ (i.e., $\mathcal{L}^{(\bar{G})}$ is one of the terms added to the original Lagrangian) we redefine $\mathcal{L}^{(\bar{G})}$ (Equation (1.68)) as

$$\mathcal{L}^{(\bar{G})\prime}(x) = g\,\frac{(-ig)^{m-1}}{d(\bar{G})} \sum_{\underline{\alpha}} \prod_{j=R+1}^{R+m} M^{(q(j))}_{\alpha_{j,1},\ldots,\alpha_{j,s(q(j))}} \int dx_{R+2}\cdots dx_{R+m}$$

$$\left\{\left[\mathcal{X}_{\epsilon,r}(G^*)_{\beta(G^*)} + \hat{\mathcal{X}}_{\epsilon}(G^*)_{\beta(G^*)}\right] : \prod_{v\in\mathcal{U}(G^*)} \psi^{(\pi(v))}_{\alpha_v}(x_{\rho(v)}) : \right\}_{x_{R+1}=x} .$$

Here $\hat{\mathcal{X}}_\epsilon(G^*)_{\beta(G^*)}$ is a new distribution of the same form (1.70) as $\mathcal{X}_{\epsilon,r}(G^*)_{\beta(G^*)}$, with the same covariance properties under Lorentz transformations; however, $\hat{\mathcal{X}}_\epsilon(G^*)$ is of course independent of $r$ and has a finite $\epsilon \to 0$ limit. The perturbation series using $\mathcal{L}^{(\bar{G})\prime}$ will now produce amplitudes renormalized as in Definition 1.19. However, the finite renormalization (1.80) in a field theory will not be completely arbitrary; rather, $\hat{\mathcal{X}}_\epsilon(U)$ will depend only on the structure of the generalized vertex $U$.

The physical interpretation of renormalization can only be mentioned briefly here. Basically, we began with a theory with certain "bare" masses and interaction strengths; but the presence of the interaction can change these quantities. This would be true even in a theory in which renormalization was unnecessary (e.g., $\phi^4$ in 2 space-time dimensions). In the case of a theory with divergences, however, the values are shifted an "infinite" amount. Correspondingly, if we wish the final particles to have finite masses and interaction strengths, we may formally assume that the original quantities were infinite and that the "infinite" counterterms we have added to the Lagrangian bring these quantities to their correct value. This "correct value" is, of course, the value determined by experiment.

Thus a certain number of experimentally determined constants must be fed into any renormalized theory. These are incorporated into the theory by the choice of finite renormalization used. (For an explicit discussion of this process in $\phi^4$ theory, see the paper of Hepp [18].) It is at this stage that the difference between "renormalizable" and "unrenormalizable" theories occurs [2]; roughly speaking, in a renormalizable theory only a finite number of experiments are needed to determine all finite renormalizations.

Section 4. THE RESULTS OF HEPP.

Hepp [18] has given a rigorous proof of the existence of the limit of $\mathcal{R}_{\epsilon,r}(G)$ when $r$ and $\epsilon$ go to zero. In Chapter III we will need a slightly stronger form of this result. In showing how to extend Hepp's work to give this stronger form we will not repeat his proofs, but only discuss the modifications which must be made.

The introduction of the parameters $\lambda_\ell$ in the following discussion seems quite arbitrary. Later chapters will motivate this introduction.

*Definition 1.20:* Let G be a Feynman graph, as in Section 3(A). For each line $\ell$, define a new propagator given by

$$\tilde{\Delta}^{(\ell)}(\lambda_\ell)(p) = Z^{(\ell)}(p) \frac{e^{\frac{1}{2}i\pi\lambda_\ell}}{(2\pi)^2} (p^2 - m^2 + i0)^{-\lambda_\ell}$$

where $Z^{(\ell)}$ is as in (1.33) and $\lambda_\ell$ is any complex number (for a discussion of the distribution $(p^2 - m^2 + i0)^{-\lambda_\ell}$, see Appendix B). We may write $\Delta^{(\ell)}(\lambda_\ell) = \lim_{\epsilon\to 0} \lim_{r\to 0} \Delta^{(\ell)}_{\epsilon,r}(\lambda_\ell)$, with

$$\tilde{\Delta}^{(\ell)}_{\epsilon,r}(\lambda_\ell)(p) = \frac{Z^{(\ell)}(p)}{(2\pi)^2} \frac{1}{\Gamma(\lambda_\ell)} \int_r^\infty a_\ell^{\lambda_\ell - 1} da_\ell \, e^{ia_\ell(p^2 - m^2 + i\epsilon)}$$

Let $\hat{\mathfrak{X}}$ be a finite renormalization (Definition 1.18), and define $\mathfrak{X}'_{\epsilon,r}(\underline{\lambda}; V_1', \ldots, V_m')$, $\bar{\mathcal{R}}'_{\epsilon,r}(\underline{\lambda}; V_1', \ldots, V_m')$, and $\mathcal{R}'_{\epsilon,r}(\underline{\lambda}; V_1', \ldots, V_m')$ as in Definition 1.19, but replacing $\Delta^{(\ell)}_{\epsilon,r}$ by $\Delta^{(\ell)}_{\epsilon,r}(\lambda_\ell)$ throughout.

Our main result is

*Theorem 1.21:* Let $\Omega = \{\underline{\lambda} \mid \text{Re}\,\lambda_\ell > 1 - \frac{1}{2L},\ \ell = 1, \ldots, L\}$. Then

$$\mathcal{R}'(\underline{\lambda}; V_1, \ldots, V_n) = \lim_{\epsilon\to 0} \lim_{r\to 0} \mathcal{R}'_{\epsilon,r}(\underline{\lambda}; V_1, \ldots, V_n) \tag{1.84}$$

exists, and is analytic, in $\Omega$. (The limit is taken in $\mathcal{S}'(R^{4n})$.)

This theorem generalizes the work of Hepp through the presence of the $\lambda_\ell$'s (and the corresponding analyticity) and through the finite renormalization used in (1.81), that is, Hepp proves the existence of the limit (1.84) for $\lambda_\ell = 1$ and for $\hat{\mathfrak{X}} = 0$. We begin our discussion by showing exactly how the finite renormalization present in (1.68) contributes to the final amplitude $\mathcal{R}'_{\epsilon,r}(\underline{\lambda}; V_1, \ldots, V_n)$.

*Definition 1.22:* If $U_1, \ldots, U_r$ are pairwise disjoint generalized vertices of G, with $\bigcup_{i=1}^r U_i = \{V_1', \ldots, V_m'\}$, the graph $G(U_1, \ldots, U_r)$ is obtained from $G(V_1', \ldots, V_m')$ by collapsing each generalized vertex $U_i$, and any lines joining two vertices of $U_i$, to a single point. The superficial divergence of $G(U_1, \ldots, U_r)$ is given by

$$\mu(U_1, \ldots, U_r) = \mu(V_1', \ldots, V_m') .$$

Note that

(1.85) $$\mu(U_1,\dots,U_r) = \Sigma'(r_\ell + 2) - 4(r-1) + \sum_{i=1}^{r} \mu(U_i) ,$$

where $\Sigma'$ runs over all lines of $G(U_1,\dots, U_r)$. Formula (1.85) should be compared with Definition 1.4.

*Definition 1.23:* Let $U_1,\dots,U_r$ be pairwise disjoint generalized vertices of G, and let $\hat{\mathcal{X}}$ be the finite renormalization of Definition 1.18. Then define recursively, for $\{U_1',\dots,U_s'\} \subset \{U_1,\dots,U_r\}$,

(1.86) $$\mathcal{X}_{\epsilon,r}(\underline{\lambda}; U_1',\dots,U_s') = \begin{cases} \hat{\mathcal{X}}_\epsilon(U_1') & \text{if } s = 1, \quad (1.86a) \\ 0 & \text{for IPR } G(U_1',\dots,U_s'), \quad (1.86b) \\ -\mathfrak{M}_{\mu(U_1',\dots,U_s')}\, \bar{\mathcal{R}}_{\epsilon,r}(\underline{\lambda}; U_1',\dots,U_s') & \text{otherwise}; \quad (1.86c) \end{cases}$$

(1.87) $$\bar{\mathcal{R}}_{\epsilon,r}(\underline{\lambda}; U_1',\dots,U_s') = \sum_P{}' \prod_{j=1}^{k(P)} \mathcal{X}_{\epsilon,r}(\underline{\lambda}; U^P_{j,1},\dots,U^P_{j,r(j)}) \prod_{\text{conn}} \Delta^{(\ell)}_{\epsilon,r}(\lambda_\ell) ;$$

(1.88) $$\mathcal{R}_{\epsilon,r}(\underline{\lambda}; U_1',\dots,U_s') = \bar{\mathcal{R}}_{\epsilon,r}(\underline{\lambda}; U_1',\dots,U_s') + \mathcal{X}_{\epsilon,r}(\underline{\lambda}; U_1',\dots,U_s') .$$

Here $\Sigma_P'$ is over all partitions P of $\{U_1',\dots,U_s'\}$ into $k(P) \geq 2$ disjoint subsets $\{U^P_{j,1},\dots,U^P_{j,r(j)}\}$, and $\prod_{\text{conn}}$ runs over all lines of $G(U_1',\dots,U_s'\}$ which connect different sets of this partition.

It is instructive to compare Definitions 1.5, 1.19, and 1.23. We see that Definition 1.23 is really similar to 1.5 rather than 1.19 because the finite renormalization enters in (1.86a) rather than in (1.86c). It is for this reason that, as we shall see below, Hepp's methods apply to the quantities of Definition 1.23 very simply. On the other hand, these are related to the quantities of Theorem 1.21 by

*Lemma 1.24:* For any $\{V_1',\dots,V_{m'}'\} \subset \{V_1,\dots,V_n\}$,

(1.89) $$\bar{\mathcal{R}}'_{\epsilon,r}(\underline{\lambda}; V_1',\dots,V_{m'}') = \sum_Q \bar{\mathcal{R}}_{\epsilon,r}(\underline{\lambda}; U_1^Q,\dots,U^Q_{s(Q)}) ,$$

(1.90) $$\mathcal{X}'_{\epsilon,r}(\underline{\lambda}; V_1',\dots,V_{m'}') = \sum_Q \mathcal{X}_{\epsilon,r}(\underline{\lambda}; U_1^Q,\dots,U^Q_{s(Q)}) ,$$

(1.91) $$\mathcal{R}'_{\epsilon,r}(\underline{\lambda}; V_1',\dots,V_{m'}') = \sum_Q \mathcal{R}_{\epsilon,r}(\underline{\lambda}; U_1^Q,\dots,U^Q_{s(Q)}) .$$

Here $\Sigma_Q$ runs over all partitions of $\{V_1',\dots,V_{m'}'\}$ into $s(Q)$ generalized vertices $U_1^Q,\dots,U^Q_{s(Q)}$.

*Proof:* The proof is by induction on $m'$; the statements are trivially true for $m' = 1$. Assume they hold for $m' < m$, and consider (1.89) with $m' = m$. By Definition 1.19 and the induction assumption,

$$\begin{aligned}
\bar{\mathcal{R}}'_{\epsilon,r}(\underline{\lambda}; V'_1, \dots, V'_m) &= \sum_P{}' \prod_{j=1}^{k(P)} \mathcal{X}(\underline{\lambda}; V^P_{j,1} \cdots V^P_{jr(j)}) \prod_{\text{conn}} \Delta^{(\ell)}_{r,\epsilon}(\lambda_\ell) \\
&= \sum_P{}' \prod_{j=1}^{k(P)} \left\{ \sum_{Q_j} \mathcal{X}(\underline{\lambda}; U^{P,Q_j}_{j,1}, \dots, U^{P,Q_j}_{j,s(Q_j)}) \right\} \prod_{\text{conn}} \Delta^{(\ell)}_{r,\epsilon}(\lambda_\ell),
\end{aligned} \tag{1.92}$$

where $\Sigma'_P$ and $\prod_{\text{conn}}$ are defined in Definition 1.5 and $\Sigma_{Q_j}$ runs over all partitions $Q_j$ of $\{V^P_{j,1} \cdots V^P_{j,r(j)}\}$ into the $s(Q_j)$ pairwise disjoint generalized vertices $\{U^{P,Q_j}_{j,i}\}$ On the other hand, we may look at the effect of $P$ and $\{Q_j\}$ differently: let $Q$ be the partition of $\{V'_1, \dots, V'_m\}$ into the $s(Q) = \Sigma_{j=1}^{k(P)} s(Q_j)$ generalized vertices $\{U^{P,Q_j}_{j,i}\}$, and let $P'$ be the partition of this *set* of generalized vertices into the $k(P')$ subsets $\{U^{P,Q_j}_{j,1}, \dots, U^{P,Q_j}_{j,s(Q_j)}\}$. Then (1.92) becomes

$$\begin{aligned}
\mathcal{R}'_{\epsilon,r}(\underline{\lambda}; V'_1, \dots, V'_m) &= \sum_Q \sum_{P'} \sum_{j=1}^{k(P')} \mathcal{X}(\underline{\lambda}; U^{P,Q_j}_{j,1}, \dots, U^{P,Q_j}_{j,s(Q_j)}) \prod_{\text{conn}} \Delta^{(\ell)}_{\epsilon,r}(\lambda_\ell) \\
&= \sum_Q \bar{\mathcal{R}}(\underline{\lambda}; U^{P,Q_1}_{1,1}, \dots, U^{P,Q_{k(P)}}_{k(P),s(Q_{k(P)})})
\end{aligned}$$

which proves (1.89).

In discussing (1.90) (with $m' = m$), we suppose first that $G(V'_1, \dots, V'_m)$ is IPR, so that the left hand side is zero. Then each term on the right hand side is also zero, as may be seen from Definition 1.23. On the other hand, if $G(V'_1, \dots, V'_m)$ is IPI, so is $G(U_1^Q, \dots, U^Q_{s(Q)})$ for any partition $Q$ of $\{V'_1, \dots, V'_m\}$. Hence, writing $U = \{V'_1, \dots, V'_m\}$,

$$\begin{aligned}
\mathcal{X}'_{\epsilon,r}(\underline{\lambda}; V'_1, \dots, V'_m) &= -\mathcal{M}_{\mu(V'_1, \dots, V'_m)} \bar{\mathcal{R}}'_{\epsilon,r}(\underline{\lambda}; V'_1, \dots, V'_m) + \hat{\mathcal{X}}_\epsilon(V'_1, \dots, V'_m) \\
&= -\mathcal{M}_{\mu(V'_1, \dots, V'_m)} \sum_Q \bar{\mathcal{R}}_{\epsilon,r}(\underline{\lambda}; U_1^Q, \dots, U^Q_{s(Q)}) + \mathcal{X}_{\epsilon,r}(\underline{\lambda}; U) \\
&= \sum_Q \mathcal{X}_{\epsilon,r}(\underline{\lambda}; U_1^Q, \dots, U^Q_{s(Q)}) .
\end{aligned}$$

Here we have used $\bar{\mathcal{R}}_{\epsilon,r}(\underline{\lambda}; U) = 0$. This proves (1.90); then (1.91) follows by adding

(1.89) and (1.90).

Theorem 1.21 now follows immediately from Lemma 1.24 and

*Lemma 1.25:* Let $W_1, \ldots, W_r$ be pairwise disjoint generalized vertices of G; let $\mathcal{L}'$ be the collection of lines of $G(W_1, \ldots, W_r)$, with $\#(\mathcal{L}') = L'$, and let $\Omega' = \{\underline{\lambda} \mid \mathrm{Re}\, \lambda_\ell > 1 - \frac{1}{2L'}, \ell \in \mathcal{L}'\}$. Then

$$\mathcal{R}(\underline{\lambda};\, W_1, \ldots, W_r) = \lim_{\epsilon \to 0} \lim_{r \to 0} \mathcal{R}_{\epsilon,r}(\underline{\lambda};\, W_1, \ldots, W_r) \tag{1.93}$$

exists, and is analytic, for $\underline{\lambda} \in \Omega'$.

*Proof:* The proof involves only slight changes of Hepp's work, and we content ourselves with giving these modifications. Throughout, $V_i$ is replaced by $W_i$, $\mathcal{L}$ by $\mathcal{L}'$, the graph $G(V_1, \ldots, V_n)$ by $G(W_1, \ldots, W_r)$, etc. In particular, "generalized vertex" now refers to a subset of $\{W_1, \ldots, W_r\}$. The combinatories of his section 2 ("Tree Structure of the $\mathcal{R}$-Operation") are otherwise unchanged; the presence of the $\lambda_\ell$'s has no effect here. The only major change is in the definition of the Feynman amplitude of a twig, which becomes, "If $U(I) = W_i$, then $\mathcal{F}_{\mathfrak{M}}(U(I)) = \hat{\mathfrak{X}}_\epsilon(W_i)$, and U(I) is called a 'twig.' "

Hepp's Lemma 3.1 must be changed to:

"Let $T = (\mathfrak{U}, \mathfrak{M}, \sigma)$ be a tree in the decomposition of $\mathcal{R}(\underline{\lambda};\, W_1, \ldots, W_r)$ in (3.1). Then the Feynman amplitude $\mathcal{F}_{\mathfrak{M}}(U(I))$ of any bough $U(I) \in \mathfrak{U}$ is for fixed $\alpha$ a finite sum of terms of the form

$$\delta\left(\sum_{W_i \in \mathcal{V}_0(I)} q_i\right) \int_0^1 \cdots \int_0^1 \left[\prod_{I' \leq I} d\tau(I')\right]\left[\prod_{\ell \in \mathcal{L}_0(I)} \alpha_\ell^{\lambda_\ell - 1}\right] P^I(p) Q^I(\alpha, \tau) \left[\prod_{I' \leq I} D^{I'} R^{I'}(A^{I'})\right.$$
$$\left. \times\, S^{I'}(B^{I'})\right] \exp\left[i \sum_{W_i, W_j \in \mathcal{V}_0(I)} \tau(I)^2 A^I_{ij}\, q_i q_j - i \sum_{\ell \in \mathcal{L}_0(I)} \alpha_\ell (m_\ell^2 - i\epsilon)\right].$$

If $\sigma(U(I)) = -1$, then $\mathcal{F}_{\mathfrak{M}}(U(I))$ has the same form except for $\Sigma\, \tau(I)^2 A_{ij}\, q_i q_j$ being replaced by zero. Here $q_i = \Sigma_{V_j \in W_i} p_j$."

The conditions 1–8 are unchanged except that in 8 we now have $x(I') = \mu(U(I'))$ for $\sigma(I') = 0$. The proof of this lemma is unchanged.

Lemma 3.2 may be replaced by

"Let $\mathcal{R}(\underline{\lambda};\, W_1, \ldots, W_r) = \Sigma_T\, \mathcal{F}_T(\underline{\lambda};\, W_1, \ldots, W_r)$ hold as above in (3.1). Suppose $\underline{\lambda} \in \Omega'$. Then the $\alpha$-integrand of every $\mathcal{F}_T(W_1, \ldots, W_r)$ is, together with all p and $\lambda$-derivatives, absolutely integrable for $r \to 0$."

This clearly implies the analyticity of the $r \to 0$ limit in (1.93) for $\underline{\lambda} \in \Omega'$. The proof is simple: not only is Hepp's expression (3.32) locally integrable for $r \to 0$, it remains so when multiplied by

$$\prod_{\ell \in \mathcal{L}_0(I)} \alpha_\ell^{\lambda_\ell - 1} ,$$

assuming $\underline{\lambda} \in \Omega'$. The $\lambda$-derivatives of this expression are also locally integrable. The proof that (3.33) is continuous and bounded when $r \to 0$ is unchanged.

This completes the discussion of the $r \to 0$ limit in (1.93). The discussion of the $\epsilon \to 0$ limit given by Hepp could be extended similarly, but we give an independent treatment. It may be seen from Hepp's Lemmas 3.2 and 3.3 (as modified) that $\lim_{r \to 0} \mathcal{R}_{\epsilon, r}(\underline{\lambda}\,;\, W_1, \ldots, W_r)$ is a sum of terms of the form

$$\delta\left(\sum_{i=1}^{r} q_i\right) \int_{0 \le \beta_{\ell_1} \le \cdots \le \beta_{\ell_L},} \left[\prod_{\ell \in \mathcal{L}'} d\beta_\ell \, \beta_\ell^{\lambda_\ell - 1}\right] \int_0^1 \cdots \int_0^1 \Pi\, d\tau(I)\, K(\beta, \tau, p)\, \delta\Big(1 - \sum_{\mathcal{L}'} \beta_\ell\Big)$$

$$\left[\sum_{i,j=1}^{r} B_{ij}(\beta, \tau)\, q_i q_j - i \sum_{\mathcal{L}'} \beta_\ell (m_\ell^2 - i\epsilon)\right]^{(k - \Sigma_{\mathcal{L}'} \lambda_\ell)}$$

Here we have set $t = \Sigma\, \alpha_\ell$, $\alpha_\ell = t\beta_\ell$, and performed an integration over $t$. The expression

$$\left[\prod_{\mathcal{L}'} \beta_\ell^{\lambda_\ell - 1}\right] K(\beta, \tau, p)$$

is integrable, $B_{i,j}(\beta, \tau)$ is continuous in $\beta$ and $\tau$, and $k$ is some integer. The existence and analyticity of the $\epsilon \to 0$ limit in (1.94) now follows directly from Theorem B.8. [Note that the integration region in (1.94) is really compact due to the factor $\delta(1 - \Sigma\, \beta_\ell)$.] This completes the proof of Lemma 1.25, and also of Theorem 1.21.

## CHAPTER II

### *Definition of Generalized Amplitudes*

Section 1. INTRODUCTION.

In Chapter I we discussed the correspondence between Feynman amplitudes and Feynman graphs. To recapitulate: if we have a graph G (see Appendix A) composed of n vertices $V_1, \ldots, V_n$ and L oriented lines (indexed by $\mathcal{L} = \{1, \ldots, L\}$), the corresponding Feynman amplitude is the distribution in $\mathcal{S}'(R^{4n})$ given formally by

$$(2.1) \qquad \mathcal{T}_G(x_1, \ldots, x_n) = \prod_{\ell \in \mathcal{L}} \Delta^{(\ell)}(x_{f_\ell} - x_{i_\ell}) \ .$$

Here $V_{f_\ell}$ and $V_{i_\ell}$ are the final and initial vertices, respectively, of the $\ell^{th}$ line, and $\Delta^{(\ell)}$ is the Feynman propagator associated with this line. It was pointed out, however, that (2.1) is not really well defined, and a definition of the renormalized Feynman amplitude – a modification of (2.1) that is well defined – was given.

Now the Feynman amplitude (2.1) as it occurred in Chapter I was a summand in a certain perturbation series. More precisely, each propagator $\Delta^{(\ell)}$ depended on two indices $\alpha_1(\ell)$ and $\alpha_2(\ell)$, and the term in the series had the form

$$(2.2) \qquad \sum_{\underline{\alpha}}{}' \ [\text{coefficient}]_{\underline{\alpha}} \prod_{\mathcal{L}} \Delta^{(\ell)}_{\alpha_1(\ell)\,\alpha_2(\ell)}(x_{f_\ell} - x_{i_\ell}) \ ,$$

the summation taken over the indices $\alpha$ corresponding to internal vertices. The corresponding graph G was given by a pair $(q, \mathcal{L})$ where q specified the types of interaction at the various internal vertices. For any two graphs having the same q (or, more generally, graphs $(q, \mathcal{L})$ and $(q', \mathcal{L}')$ for which there is a permutation s of the internal vertices with $q' = q \circ s$) the set of propagators which occurs in the amplitude is the same. What varies is the position of the line $\ell$, that is, the values of $V_{i_\ell}$ and $V_{f_\ell}$, and the values of the coefficient in (2.2).

This suggests the possibility we explore in this chapter. We will define a generalization of the Feynman amplitude (2.1) which depends on certain parameters. For isolated values of these parameters, the generalized amplitude reduces to the amplitude for various actual graphs; thus we will provide an interpolation of the amplitudes between different graphs. In particular, using the notation of Chapter I, one generalized amplitude can interpolate between all graphs

$\{(q', \mathcal{L}') \in \mathcal{G}(\Theta, \delta, m) | q' = qs\}$ for some fixed $q$ (here $s$ is any permutation of the internal vertices $\{R+1, \ldots, R+m\}$). Chapter IV will discuss applications of this interpolation.

The Feynman amplitudes defined here are also generalized to depend on certain parameters $\lambda_1, \ldots, \lambda_L$ (not related to the parameters mentioned above). The presence of these parameters enables us to handle the divergence difficulties encountered in defining (2.1). Thus our generalized amplitude will be an analytic function of the $\lambda$'s which is formally equal to (2.1) (or more precisely, in view of the last paragraph, formally equal to the interpolation of (2.1) between graphs) when $\lambda_1 = \lambda_2 = \cdots = \lambda_L = 1$. The original divergence difficulties will now appear as a singularity of the generalized amplitude at this point. At first sight this seems to be no improvement; however, in Chapter III we discuss a method of using the analyticity in the $\lambda$'s to give a new definition of the renormalized amplitude.

Section 2. GENERALIZATION OF THE PROPAGATOR.

We begin by introducing some notation which will hold throughout this chapter. For 4-vectors $x$ and $y$ we write, as usual,

$$x^2 = x^\mu g_{\mu\nu} x^\nu$$
$$x \cdot y = x^\mu g_{\mu\nu} y^\nu \ .$$

We will be working in the space $R^{4n}$ of n-tuples of the 4-vectors; such an n-tuple is written $\underline{x} = (x_1, \ldots, x_n)$. We define a *non-contravariant* inner product in this space by

$$\underline{x} \circ \underline{y} = \sum_{i=1}^{n} \sum_{\mu=0}^{3} x_i^\mu y_i^\mu \ ,$$

(and write similarly $x \circ y = \Sigma_{\mu=0}^3 x^\mu y^\mu$ for 4-vectors). Let $G$ be the $4n \times 4n$ quadratic form $G_{i\mu,j\nu} = \delta_{ij} g_{\mu\nu}$, then the usual covariant inner product on $R^{4n}$ is written

$$\underline{x} \cdot \underline{y} = \underline{x} \circ G \circ \underline{y} \ .$$

All Fourier transforms will be taken with respect to the quadratic form $\underline{x} \cdot \underline{y}$ (or $x \cdot y$ in 4 dimensions); see Definition B.2.

We use the symbol $\otimes$ to combine vectors or matrices as $R^4$ with those on $R^n$. For example, the quadratic form $G$ may be written $G = g \otimes I$, with $I$ the n-dimensional identity matrix. In a slight abuse of notation we will sometimes combine vectors and matrices in this way; thus, for example, if $v \in R^n$, the expression $(v \otimes g) \circ \underline{x}$ denotes the 4-vector $[(v \otimes g) \circ \underline{x}]_\mu = \Sigma_{i,\nu} v_i g_{\mu\nu} x_i^\nu$.

We will frequently use other quadratic forms on $R^{4n}$; these will be written with script letters and, as above,

$$\underline{x} \circ \mathcal{P} \circ \underline{y} = \sum_{i,\mu;j,\nu} x_i^{\mu} \mathcal{P}_{i\mu,j\nu} y_j^{\nu}$$

On the other hand, we use Roman letters to denote $n \times n$-quadratic forms, and, for P such a form, define

$$\underline{x} \cdot P \cdot \underline{y} = x \circ [P \otimes g] \circ y \ .$$

*Definition 2.1:* Let a be a 4-vector, and set $\underline{a} = (a, a, \ldots, a)$. A quadratic form $\mathcal{P}$ (or P) is *translation invariant* if $\underline{x} \circ \mathcal{P} \circ (\underline{y} + \underline{a}) = \underline{x} \circ \mathcal{P} \circ \underline{y}$ [or $\underline{x} \cdot P \cdot (\underline{y} + \underline{a}) = \underline{x} \cdot P \cdot \underline{y}$] for all $\underline{x}$, $\underline{y}$, and a; equivalently, if $\Sigma_{i=1}^{n} \mathcal{P}_{i\mu,j\nu} = 0$ for all $\mu, j, \nu$ (or $\Sigma_{i=1}^{n} P_{ij} = 0$ for all j). Define $E = \{\underline{x} \in R^{4n} \mid \Sigma_{i=1}^{n} x_i^{\mu} = 0$, for all $\mu\}$. E is the orthogonal complement (with respect to either $x \cdot y$ or $x \circ y$) in $R^{4n}$ of the space of all vectors $\underline{a} = (a, a, \ldots, a)$. As in Remark B.5, we have

$$\delta_E = \prod_{\mu=0}^{4} \sqrt{n}\, \delta\Big( \sum_{i=1}^{n} x_i^{\mu} \Big) = n^2 \delta\Big( \sum_{i=1}^{n} x_i \Big) \ .$$

Finally, for any translation invariant quadratic form $\mathcal{P}$, we let $\mathcal{P}_E$ denote the restriction of $\mathcal{P}$ to E.

We now turn to the generalization of the propagator. Recall that the Feynman propagator $\Delta^{(\ell)}$ occurring in (2.1) is given in p-space by

$$\tilde{\Delta}^{(\ell)}(p) = \frac{i}{(2\pi)^2} Z_\ell(p)\, [p^2 - m_\ell^2 + i0]^{-1} \ , \tag{2.3}$$

where $Z_\ell$ is a polynomial of degree $r_\ell$. From Theorem B.11 we have

$$\Delta^{(\ell)}(x) = \frac{m_\ell}{(2\pi)^2} Z_\ell \Big(-i g \circ \frac{\partial}{\partial x}\Big) \frac{K_1[m_\ell(-x^2 + i0)^{1/2}]}{(-x^2 + i0)^{1/2}} \ . \tag{2.4}$$

*Remark 2.2:* As an illustration we calculate the factor $[\det(-g)]^{1/2}$ needed in applying Theorem B.11 to (2.3), using part (iv) of the theorem. Let $g_\epsilon(t)$ be the matrix

$$g_\epsilon(t) = \begin{bmatrix} t + i\epsilon & 0 & 0 & 0 \\ 0 & -1 + i\epsilon & 0 & 0 \\ 0 & 0 & -1 + i\epsilon & 0 \\ 0 & 0 & 0 & -1 + i\epsilon \end{bmatrix} ;$$

then according to (iv), if we define $f_\epsilon(t) = \{\det[-g_\epsilon(t)]\}^{1/2}$ to be continuous in t and satisfy $\lim_{\epsilon \to 0} f_\epsilon(-1) = 1$, we will have $\lim_{\epsilon \to 0} f_\epsilon(1) = [\det(-g)]^{1/2}$. But to first order in $\epsilon$,

$$\det[-g_\epsilon(t)] = -t + 3i\epsilon t - i\epsilon \ ;$$

defining the square root to be continuous in t then gives

$$[\det(-g)]^{1/2} = f_\epsilon(1) = -i \ .$$

We now consider a graph G as in the introduction, and let e be the incidence matrix (Definition A.6). Then, using (2.4), (2.1) may be written

$$
\begin{aligned}
\mathcal{T}_G(\underline{x}) &= \prod_{\ell \in \mathcal{L}} \Delta^{(\ell)}\left(\sum_{i=1}^{n} e_i^{(\ell)} x_i\right) \\
&= \prod_{\ell \in \mathcal{L}} \frac{m_\ell}{(2\pi)^2} Z_{f'}\left[-\frac{i}{2}(e^{(\ell)} \otimes g) \circ \frac{\partial}{\partial \underline{x}}\right] \\
&\quad \times \frac{K_1[m_\ell(\underline{x} \cdot Q^{(\ell)} \cdot \underline{x} + i0)^{1/2}]}{[\underline{x} \cdot Q^{(\ell)} \cdot \underline{x} + i0]^{1/2}} ,
\end{aligned} \tag{2.5}
$$

where $Q^{(\ell)}$ is the quadratic form

$$
Q_{ij}^{(\ell)} = -e_i^{(\ell)} e_j^{(\ell)} \tag{2.6}
$$

and, as remarked before $(e^{(\ell)} \otimes g) \circ \frac{\partial}{\partial \underline{x}} = \Sigma_{i,\nu}\, e_i^{(\ell)} g_{\mu\nu} \frac{\partial}{\partial x_i^\nu}$. This suggests immediately a generalization of the propagator: the introduction of more general quadratic forms than those of the form (2.6). It would be possible to do this by introducing a new matrix e and requiring that (2.6) still hold, but for convenience we shall assume no relation between $e^{(\ell)}$ and $Q^{(\ell)}$ in the following definition. We also introduce a parameter $\lambda$ for reasons discussed in the introduction; the rather complicated form in which it appears is discussed in Remark 2.5 (c).

*Definition 2.3:* Let Q be a translation invariant real $n \times n$ quadratic form, e a vector in $R^n$ satisfying $\Sigma_{i=1}^n e_i = 0$, $\lambda$ a complex number, m a positive real number, and Z a polynomial of degree r in $C[x^0, x^1, x^2, x^3]$. Then the *generalized Feynman propagator* (GFP) $\Delta(\lambda, Q, e, Z, m) \in \mathcal{S}'(R^{4n})$ is given by

$$
\Delta(\lambda, Q, e, Z, m) = a(\lambda)\, Z[-\tfrac{i}{2}(e \otimes g) \circ \frac{\partial}{\partial \underline{x}}]\, m^{2-\lambda} \times \frac{K_{2-\lambda}[m[\mathcal{Q}_1 + i0]^{1/2}]}{[(\mathcal{Q}_1 + i0)^{1/2}]^{2-\lambda}} ,
$$

with $a(\lambda) = \dfrac{i\, e^{-\frac{1}{2}\lambda\pi i}}{2^\lambda \Gamma(\lambda)\, 2\pi^2}$, and $\mathcal{Q}_1 = Q \otimes g$. With some abuse of notation we write

$$
\Delta(\lambda, Q, e, Z, m) = \lim_{\mathcal{Q}_2 \to 0} \Delta(\lambda, \mathcal{Q}, e, Z, m) .
$$

Here $\Delta(\lambda, \mathcal{Q}, e, Z, m)$ is defined as the right hand side of (2.7) with $\mathcal{Q}_1$ replaced by $\mathcal{Q} = \mathcal{Q}_1 + i\mathcal{Q}_2$ throughout, and $\mathcal{Q}_2$ is a translation invariant quadratic form which is positive definite on $E[(\mathcal{Q}_2)_E > 0]$.

It is now natural to make

*Definition 2.4:* Let $\underline{Q}$, $\underline{e}$, $\underline{\lambda}$, $\underline{m}$, and $\underline{Z}$ be L-tuples of the corresponding quantities of Definition 2.3. The *generalized Feynman amplitude* (GFA) $\mathcal{T}(\underline{\lambda}, \underline{Q}, \underline{e}, \underline{Z}, \underline{m})$ is defined *formally* by

(2.8) $$\mathcal{T}(\underline{\lambda}, \underline{Q}, \underline{e}, \underline{Z}, \underline{m}) = \prod_{\ell=1}^{L} \Delta(\lambda_\ell, Q^{(\ell)}, e^{(\ell)}, Z_\ell, m_\ell) .$$

*Remark 2.5:* (a). It must be emphasized that Definition 2.4 is only formal. We will proceed to make it precise by the following steps:

(i) Define $\mathcal{T}(\underline{\lambda}, \underline{\mathcal{Q}}, \underline{e}, \underline{Z}, \underline{m})$ precisely, for $\underline{\lambda}$ subject to certain restrictions, by

$$\mathcal{T}(\underline{\lambda}, \underline{\mathcal{Q}}, \underline{e}, \underline{Z}, \underline{m}) = \prod_{1}^{L} \Delta(\lambda_\ell, \mathcal{Q}^{(\ell)}, e^{(\ell)}, Z_\ell, m_\ell) .$$

(ii) Show the existence of the limit

$$\mathcal{T}(\underline{\lambda}, \underline{Q}, \underline{e}, \underline{Z}, \underline{m}) = \lim_{\mathcal{Q}_2^{(\ell)} \to 0} \mathcal{T}(\underline{\lambda}, \underline{\mathcal{Q}}, \underline{e}, \underline{Z}, \underline{m})$$

under even more stringent limitations on $\underline{\lambda}$, and some restrictions on $\underline{Q}$.

(iii) Extend the definition in (ii) to all $\underline{\lambda} \in \mathbb{C}^L$ by analytic continuation.

(b). Notice that if $e_r^{(\ell)}$ is the incidence matrix of a Feynman graph, $Q^{(\ell)}$ is given by (2.6), and $\lambda_\ell = 1$ for all $\ell$, then (2.8) reduces to (2.5), the Feynman amplitude of the graph. Thus by keeping $\lambda_\ell = 1$ and varying $\underline{Q}$ and $\underline{e}$ we can (formally) vary $\mathcal{T}$ between the Feynman amplitudes for various graphs.

(c). The motivation for introducing the complicated function $a(\lambda)$ in Definition 2.3 was to make this definition of the GFP agree with that in [33].

(d). None of the results of this thesis would be changed by introducing in (2.7), instead of $Z(y)$, a new polynomial (in $y$) $Z(\lambda, y)$, of the same degree, whose coefficients are analytic in $\lambda$ and satisfy $Z(1, y) = Z(y)$. This would affect the finite renormalization constant appearing in Chapter III, Section 4.

(e). We frequently suppress the dependence of $\mathcal{T}$ on $\underline{Z}$ and $\underline{m}$, which is not of primary concern to us.

## Section 3. PRELIMINARY DEFINITION OF THE GFA.

In this section we discuss part (a.i) of Remark 2.5. The basic result is

*Lemma 2.6:* Let $\mathcal{Q} = \mathcal{Q}_1 + i\mathcal{Q}_2$ be a translation invariant quadratic form with $\mathcal{Q}_2$ positive definite on E. Then $\Delta(\lambda, \mathcal{Q}, e, Z, m)$ is an $L^\infty$ function on $R^n$ whenever $\mathrm{Re}\,\lambda > \frac{1}{2}(r+4)$ (here $r$ is the degree of $Z$; see Definition 2.3).

*Proof:* Because $\mathcal{Q}$ is translation invariant, the distribution $\Delta(\lambda, \mathcal{Q}, e, Z, m)$ is really of the form

$$I \otimes \Delta_E(\lambda, \mathcal{Q}, e, Z, m) ,$$

where I is the identity on the orthogonal complement of E, and $\Delta_E$ is $\Delta$ restricted to E. Thus, from (2.7), (B.3), and (B.11), we have

$$\tilde{\Delta}(\lambda, \mathcal{Q}, e, Z, m) = \frac{2^{2n-4}\, i\, e^{\frac{1}{2}\pi i\lambda}}{(2\pi)^2} \frac{\Gamma(\lambda + 2n - 4)}{\Gamma(\lambda)} [\det \mathcal{P}_E]^{\frac{1}{2}}$$

(2.9)
$$\times [(2\pi)^2 \delta_E] Z(\tfrac{1}{2}\, e \circ p)(\mathcal{P} - m^2)^{4-2n-\lambda}$$

$$= (2\pi)^2 \delta_E\, \tilde{\Delta}'(\lambda, \mathcal{Q}, e, Z, m) ,$$

where $\mathcal{P} = -G \circ \mathcal{Q}_E^{-1} \circ G$, and $\det \mathcal{P}_E = \det - \mathcal{P}_E$ because E has even dimension. But for $\operatorname{Re}\lambda > \frac{1}{2}(r+4)$, $\tilde{\Delta}'$ is a continuous function in $L^1(E)$, by simple power counting. Since the Fourier transform of an $L^1$ function is $L^\infty$,

(2.10)
$$\Delta(\lambda, \mathcal{Q}, e, Z, m) = I \otimes \mathcal{F}^{-1}[\tilde{\Delta}'(\lambda, \mathcal{Q}, e, Z, m)]$$

is $L^\infty$ on $R^n$.

*Definition 2.7:* Let $\mathcal{Q}^{(1)}, \ldots, \mathcal{Q}^{(L)}$ be translation invariant quadratic forms with $\mathcal{Q}_2^{(\ell)}$ positive definite on E. Then for $\operatorname{Re}\lambda_\ell > \frac{1}{2}(r_\ell + 4)$, define

$$\mathcal{T}(\underline{\lambda}, \underline{\mathcal{Q}}, \underline{e}, \underline{Z}, \underline{m}) = \prod_{\ell \in \mathcal{L}} \Delta(\lambda_\ell, \mathcal{Q}^{(\ell)}, e^{(\ell)}, Z_\ell, m_\ell) .$$

We remark that this is justified by Lemma 2.6. It is, in fact, our first rigorous definition of a GFA.

We now calculate $\mathcal{T}$ using (2.9), (2.10), and two easily verified formulas which we give in

*Lemma 2.8:* (a). Suppose $\operatorname{Re}\nu > 0$ and $\operatorname{Im}\gamma > 0$. Then

$$\gamma^{-\nu} = \frac{e^{-\nu\pi i/2}}{\Gamma(\nu)} \int_0^\infty d\alpha\, \alpha^{\nu-1} \exp i\alpha\gamma .$$

(b). Suppose $\mathcal{R}$ is a quadratic form on $R^m$, with $\operatorname{Re}\mathcal{R} > 0$. Then for any $\underline{B} \in C^m$,

$$\int_{R^m} d\underline{y} \exp[-\underline{y} \circ \mathcal{R} \circ \underline{y} + \underline{y} \circ \underline{B}] = \frac{\pi^{m/2}}{[\det \mathcal{R}]^{\frac{1}{2}}} \exp[\tfrac{1}{4} B \circ \mathcal{R}^{-1} \circ B] .$$

Here the sign of $[\det \mathcal{R}]^{\frac{1}{2}}$ is determined by analytic continuation from the region where $\mathcal{R}$ is real and positive definite, in which $[\det \mathcal{R}]^{\frac{1}{2}} > 0$.

*Proof:* (a) is trivial; (b) is proved by diagonalizing $\mathcal{R}$.

We now calculate the inverse Fourier transform of $\tilde{\Delta}'$ (see 2.10), using Lemma 2.8 (a) to express $(\mathcal{P} - m^2)^{4-2n-\lambda}$ as an $\alpha$-integral. Let $d\mu$ be the Lebesgue measure on E. Then, writing

$$Z(\tfrac{1}{2} e \circ \underline{p}) = [Z(\tfrac{i}{2} e \circ G \circ \frac{\partial}{\partial \underline{s}}) e^{-i\underline{p} \circ G \circ \underline{s}}]_{s=0} ,$$

we have

$$\mathcal{F}^{-1}[\tilde{\Delta}'(\lambda, \mathcal{Q}, e, Z, m)](\underline{x}) = \frac{(-1)^n 2^{2n-4} i}{(2\pi)^2 \Gamma(\lambda)} [\det \mathcal{P}_E]^{1/2} \frac{1}{(2\pi)^{2n-2}} \int_E d\mu(\underline{p})$$

(2.11)

$$\times \int_0^\infty d\alpha \, \alpha^{\lambda+2n-5} Z(\tfrac{i}{2} e \circ G \circ \frac{\partial}{\partial \underline{s}}) \exp i[\alpha \underline{p} \circ \mathcal{P} \circ \underline{p} + \underline{p} \circ G \circ (\underline{x} - \underline{s}) - \alpha m^2]\Big|_{s=0} .$$

Now apply Lemma 2.8 (b) with $\mathcal{R} = -i\alpha \mathcal{P}_E$. This will produce a factor $[\det \mathcal{R}]^{-1/2}$, but this does *not* necessarily correspond to the factor $[\det \mathcal{P}_E]^{1/2}$ from (2.9), due to a possible difference in the branch of the square root. We compare the two values for $\mathcal{R} = -i\alpha\mathcal{P}_E = 1_E$ (the identity on E), where by Lemma 2.8, $[\det \mathcal{R}]^{1/2} = 1$. Let $\mathcal{P}_E(\theta) = -\alpha e^{-i\theta} 1_E$, for $0 \le \theta \le \pi/2$. Then $\det \mathcal{P}_E(\theta) = \alpha^{4(n-1)} e^{-4(n-1)i\theta}$, and since (iv) of Theorem B.11 implies $[\det \mathcal{P}_E(0)]^{1/2} > 0$, we have in general $[\det \mathcal{P}_E(\theta)]^{1/2} = \alpha^{2n-2} e^{-2(n-1)i\theta}$. Thus for $\theta = \pi/2$, where $-i\alpha\mathcal{P}_E(\theta) = 1_E$, we have $[\det \mathcal{P}_E(\pi/2)]^{1/2} = \alpha^{2n-2} e^{-(n-1)\pi i}$. But we already found that at this point $[\det \mathcal{R}]^{1/2} = 1$, so in general,

(2.12) $$[\det \mathcal{R}]^{1/2} = (-1)^{n-1} \alpha_\ell^{2n-2} [\det \mathcal{P}_E]^{1/2}$$

Thus from (2.11), (2.12), and Lemma 2.8 (b),

$$\mathcal{F}^{-1}[\tilde{\Delta}'(\lambda, \mathcal{Q}, e, Z, m)](\underline{x}) = \frac{-i}{4(2\pi)^2 \Gamma(\lambda)} \int_0^\infty d\alpha \, \alpha^{\lambda-3}$$

(2.13)

$$\times Z(\tfrac{i}{2} e \circ G \circ \frac{\partial}{\partial \underline{s}}) \exp i[\tfrac{1}{4} (\underline{x} - \underline{s}) \circ \frac{\mathcal{Q}}{\alpha} \circ (\underline{x} - \underline{s}) - \alpha m^2]\Big|_{s=0} .$$

From (2.10) and Definition 2.7, we have

(2.14) $$\mathcal{T}(\underline{\lambda}, \underline{\mathcal{Q}}, \underline{e}, \underline{Z}, \underline{m}) = I \otimes \prod_{\ell} \mathcal{F}^{-1} \tilde{\Delta}'(\lambda_\ell, \mathcal{Q}^{(\ell)}, e^{(\ell)}, Z_\ell, m_\ell) .$$

We take the Fourier transform of (2.14), using (2.13) and (B.3):

$$\tilde{\mathcal{T}}(\underline{\lambda}, \underline{\mathcal{Q}}, \underline{e}, \underline{Z}, \underline{m}) = (2\pi)^2 \delta_E \int_E \frac{d\mu(\underline{x})}{(2\pi)^{2n-2}} \int_0^\infty \cdots \int_0^\infty d\alpha_1 \ldots d\alpha_L$$

(2.15)

$$\times \prod_{\ell} \left\{ \alpha_\ell^{\lambda_\ell - 3} \left[ \frac{-i}{4(2\pi)^2 \Gamma(\lambda_\ell)} \right] Z_\ell(\tfrac{i}{2} e^\ell \circ G \circ \frac{\partial}{\partial \underline{s}_\ell}) \right\}$$

$$\times \exp i \left\{ -\tfrac{1}{4} \underline{x} \circ \mathcal{A} \circ \underline{x} - \underline{x} \circ [G \circ \underline{p} + \tfrac{1}{2} \sum_{\ell} \frac{\mathcal{Q}^{(\ell)}}{\alpha_\ell} \circ \underline{s}^{(\ell)}] + \sum_{\ell} \left[ \underline{s}^{(\ell)} \circ \frac{\mathcal{Q}^{(\ell)}}{4\alpha_\ell} \circ \underline{s}^{(\ell)} - \alpha_\ell m_\ell^2 \right] \right\} \Big|_{\underline{s}^{(\ell)} = 0} .$$

Here we have put

$$\mathcal{A} = -\sum_{\mathcal{L}} \frac{\mathcal{Q}^{(\ell)}}{a_\ell} . \tag{2.16}$$

We do the $\underline{x}$ integral in (2.15) by another application of Lemma 2.8 (b). Thus we have finally

$$\begin{aligned}
\tilde{\mathcal{J}}(\underline{\lambda}, \underline{\mathcal{Q}}, \underline{e}, \underline{Z}, \underline{m}) &= b(\underline{\lambda})\, \delta_E \int_0^\infty \cdots \int_0^\infty da_1 \cdots da_L \prod_{\mathcal{L}} \{ a_\ell^{\lambda_\ell - 3} \\
&\times Z_\ell\left(\tfrac{i}{2}\, e^{(\ell)} \circ G \circ \frac{\partial}{\partial \underline{s}^{(\ell)}}\right)\} \, [\det \mathcal{A}_E]^{-1/2} \exp i \{ \underline{p} \circ G \circ \mathcal{A}_E^{-1} \circ G \circ \underline{p} \\
&+ \sum_{\mathcal{L}} \underline{s}^{(\ell)} \circ \frac{\mathcal{Q}^{(\ell)}}{a_\ell} \circ \mathcal{A}_E^{-1} \circ G \circ \underline{p} + \frac{1}{4} \sum_{\mathcal{L} \times \mathcal{L}} \frac{\underline{s}^{(\ell)} \circ \mathcal{Q}^{(\ell)} \circ \mathcal{A}_E^{-1} \circ \mathcal{Q}^{(\ell')} \circ \underline{s}^{(\ell')}}{a_\ell a_{\ell'}} \\
&+ \sum_{\mathcal{L}} \left(\underline{s}^{(\ell)} \circ \frac{\mathcal{Q}^{(\ell)}}{4a_\ell} \circ \underline{s}^{(\ell)} - a_\ell m_\ell^2\right) \} \Big|_{\underline{s}^{(\ell)} = 0} ,
\end{aligned} \tag{2.17}$$

where we have collected various constants in

$$b(\underline{\lambda}) = \frac{2^{2n-4L} \pi^{2n-2L-2} (-i)^{2n-2+L}}{\prod_{\mathcal{L}} \Gamma(\lambda_\ell)} . \tag{2.18}$$

Formula (2.17) will be our starting point for further investigations of the GFA.

## Section 4. Reduction of spin terms.

Before discussing the $\mathcal{Q}_2^{(\ell)} \to 0$ limit of (2.17) we present a notation for writing the polynomials arising from spin and derivative coupling. Suppose we have an index set $I$, for each $i \in I$ an indeterminant $X_i$, and for each $(i,j) \in I \times I$ an indeterminant $Y_{ij}$. For historical reasons $Y_{ij}$ is called the *contraction* of $X_i$ and $X_j$ and is sometimes written $Y_{ij} = \overline{X_i X_j}$. $C[X]$ denotes polynomials in $\{X_i\}$ over $C$, similarly $C[X,Y]$.

*Definition 2.9:* The *$\tau$-product* is the $C$-linear mapping $\tau\colon C[X] \to C[X,Y]$ defined on monomials by

$$\tau(X_{i_1}, \dots, X_{i_L}) = \sum_{m=0}^{[L/2]} \sum_{\substack{j_1 \cdots j_m \\ k_1 \cdots k_m}} X_{i_1} \cdots \hat{X}_{j_1} \cdots \hat{X}_{k_m} \cdots X_{i_L} \; Y_{j_1 k_1} \cdots Y_{j_m k_m} \tag{2.19}$$

Here the second sum is over all $\{j_1, \dots, k_m\}$ satisfying $j_1 < j_2 < \cdots < j_m$, $k_1 < k_2 < \cdots < k_m$, and $j_r < k_r$ for $1 \le r \le m$.

The $\tau$-product is closely related to Wick's T-product of quantum fields [39; see also Definition 1.2]. Specifically, (2.19) corresponds directly to Wick's theorem for the expansion of a T-product of boson fields in terms of contractions and normal products, with the normal product of field operators corresponding to the ordinary product of the $X_i$'s.

Definition 2.9 may also be written in terms of Caianello's hafnian [5] as

$$\tau(X_{i_1}\cdots X_{i_L}) = \sum_{m=0}^{[L/2]} \sum_{\{j_1\cdots j_m, k_1\cdots k_m\}\subset\{1\cdots L\}} X_{i_1}\cdots \hat{X}_{j_1}\cdots \hat{X}_{k_m}\cdots X_{i_L}\,[j_1 j_2\cdots k_m],$$

where $[j_1,\ldots,k_m]$ is the hafnian defined with $[ij] = Y_{ij}$.

Now the part of (2.17) that we wish to study in this section is

$$\prod_{\mathcal{L}} Z_\ell\left(\tfrac{i}{2}\, e^{(\ell)}\circ G\circ \frac{\partial}{\partial \underline{s}^{(\ell)}}\right) \exp i\left\{\sum_{\mathcal{L}} \underline{s}^{(\ell)}\circ \frac{\mathcal{Q}^{(\ell)}}{a_\ell}\circ \mathcal{A}_E^{-1}\circ G\circ \underline{p}\right.$$

(2.20)

$$\left. + \frac{1}{4}\sum_{\mathcal{L}\times\mathcal{L}} \underline{s}^{(\ell)}\circ\left[\frac{\mathcal{Q}^{(\ell)}}{a_\ell}\circ \mathcal{A}_E^{-1}\circ \frac{\mathcal{Q}^{(\ell')}}{a_{\ell'}} + \delta_{\ell\ell'}\frac{\mathcal{Q}^{(\ell)}}{a_\ell}\right]\circ \underline{s}^{(\ell')}\right\}\Bigg|_{\underline{s}^{(\ell)}=0}.$$

*Lemma 2.10:* Define

(2.21a) $$X^{(\ell)} = -\frac{1}{2}\,[e^{(\ell)}\otimes g]\circ \frac{\mathcal{Q}^{(\ell)}}{a_\ell}\circ \mathcal{A}_E^{-1}\circ G\circ \underline{p}\quad ,$$

(2.21b) $$\overline{X^{(\ell)} X^{(\ell')}} = -\frac{i}{8}\,[e^{(\ell)}\otimes g]\circ\left[\frac{\mathcal{Q}^{(\ell)}}{a_\ell}\circ \mathcal{A}_E^{-1}\circ \frac{\mathcal{Q}^{(\ell')}}{a_{\ell'}} + \delta_{\ell\ell'}\frac{\mathcal{Q}^{(\ell)}}{a_\ell}\right]\circ [e^{(\ell')}\otimes g].$$

Then (2.20) is given by

(2.22) $$\tau\Big[\prod_{\ell\in\mathcal{L}} Z_\ell(X^{(\ell)})\Big]\,.$$

*Remark 2.11:* Note that $X^{(\ell)}$ is really a 4-vector, with

$$X^{(\ell)\mu} = -\frac{1}{2a_\ell}\sum_{i_1,i_2,i_3=1}^{n}\ \sum_{\nu_1,\nu_2,\nu_3,\nu_4=0}^{4} g^{\mu\nu_1}\, e^{(\ell)}_{i_1}\, \mathcal{Q}^{(\ell)}_{i_1\nu_1,i_2\nu_2}\, (\mathcal{A}_E^{-1})_{i_2\nu_2,i_3\nu_3}\, g_{\nu_3\nu_4}\, p^{\nu_4}_{i_3}\ .$$

Thus it is a suitable argument for $Z_\ell$. Similarly, $\overline{X^{(\ell)}X^{(\ell')}}$ is a $4\times 4$ matrix $(\overline{X^{(\ell)}X^{(\ell')}})^{\mu\nu}$.

*Proof of Lemma 2.10:* The exponent appearing in (2.20) has the form

$$\exp\left[\sum_{\mathcal{L}} \underline{s}^{(\ell)}\circ\mathcal{B} + \sum_{\mathcal{L}\times\mathcal{L}} \underline{s}^{(\ell)}\circ \mathcal{C}^{(\ell,\ell')}\circ \underline{s}^{(\ell')}\right]$$

Set

$$\mathcal{D}^{(\ell)\mu}_i = [\mathcal{B}_{i\mu} + 2\sum_{j,\nu} \mathcal{C}^{(\ell,\ell')}_{i\mu,j\nu}\, s^{(\ell')\nu}_j]\ ,$$

$$\overline{\mathcal{D}^{(\ell)\mu}_i\, \mathcal{D}^{(\ell')\nu}_j} \doteq 2\,\mathcal{C}^{(\ell,\ell')}_{i\mu,j\nu}\ .$$

Then it is easily verified by induction that

$$\frac{\partial}{\partial s_{i_1}^{(\ell_1)\nu_1}} \cdots \frac{\partial}{\partial s_{i_k}^{(\ell_k)\nu_k}} \exp[\Sigma\, \underline{s}^{(\ell)} \circ \mathcal{B} + \Sigma\, \underline{s}^{(\ell)} \circ \mathcal{C}^{(\ell,\ell')} \circ \underline{s}^{(\ell')}]$$

$$= \tau\,(\mathcal{D}_{i_1}^{(\ell_1)\nu_1} \cdots \mathcal{D}_{i_k}^{(\ell_k)\nu_k})\, \exp[\Sigma\, \underline{s}^{(\ell)} \circ \mathcal{B} + \Sigma\, \underline{s}^{(\ell)} \circ \mathcal{C}^{(\ell,\ell')} \circ \underline{s}^{(\ell')}]\ .$$

Setting $\underline{s}^{(\ell)} = 0$ in this formula, then contracting (over $i_1, \ldots, i_k, \nu_1, \ldots, \nu_k$) with

$$(\tfrac{i}{2}\, e_{i_1}^{(\ell_1)}\, g_{\mu_1\nu_1}) \cdots (\tfrac{i}{2}\, e_{i_k}^{(\ell_k)}\, g_{\mu_k\nu_k})$$

gives $\tau(X^{(\ell_1)\mu_1} \cdots X^{(\ell_k)\mu_k})$, and this proves the lemma.

Formula (2.22) is the simplified form of the spin and derivative coupling contribution to (2.17) that we wished to obtain. It looks deceptively innocent, since it really contains the complicated formulas (2.21). We discuss at the end of this chapter how these formulas may be simplified by a redefinition of the GFA.

## Section 5. THE $\underline{\mathcal{Q}}_2 \to 0$ LIMIT.

In this section we turn to part (a)(ii) of Remark 2.5. Throughout, we will take $\underline{Q}$ to be an L-tuple of $n \times n$ translation invariant quadratic forms, and $\underline{\mathcal{Q}}$ the corresponding $4n \times 4n$ forms with $\mathcal{Q}_1^{(\ell)} = g \otimes Q^{(\ell)}$, $\mathcal{Q}_2^{(\ell)}$ positive definite on E. In Section 3 we defined $\mathcal{T}(\underline{\lambda}, \underline{\mathcal{Q}}, \underline{e})$; here we prove the existence of the limit

$$\mathcal{T}(\underline{\lambda}, \underline{Q}, \underline{e}) = \lim_{\mathcal{Q}_2^{(\ell)} \to 0} \mathcal{T}(\underline{\lambda}, \underline{\mathcal{Q}}, \underline{e}) \ , \tag{2.23}$$

under certain restrictions on $\underline{\lambda}$, $\underline{Q}$, and $\underline{\mathcal{Q}}_2$. This will complete our definition of the GFA (2.8) for these values of $\underline{\lambda}$; in later chapters we will extend the definition to all $\underline{\lambda}$ by analytic continuation.

We will define the GFA (2.8) only for $\underline{Q}$ satisfying two conditions (in addition to translation invariance):

(Q 1). For all $1 \leq \ell \leq L$ and $1 \leq i, j \leq n$ with $i \neq j$,

$$Q_{ij}^{(\ell)} \geq 0\ .$$

(Q 2). For all $0 < a_\ell < \infty$, the matrix

$$A = -\sum_{\ell} \frac{Q^{(\ell)}}{a_\ell} \tag{2.24}$$

has rank $n-1$.

The reader should compare (2.24) and (2.16). Note that when $e_i^{(\ell)}$ is the incidence matrix of some graph and $Q_{ij}^{(\ell)} = -e_i^{(\ell)} e_j^{(\ell)}$, condition (Q 1) is satisfied. We wish to vary $\underline{Q}$ between quadratic forms of this type; (Q 1) does not restrict our ability to do this. Condition (Q 2) could be given many equivalent forms; for example, it suffices to require:

(Q 2)′. The matrix $\Sigma_{\mathcal{L}} Q^{(\ell)}$ has rank $n-1$.

The significance of this condition shows up more clearly in another equivalent formulation which we now discuss.

Let $\Gamma = \{(\ell, i, j) | \ell \in \mathcal{L},\ i > j,$ and $Q_{ij}^{(\ell)} \neq 0\}$. We construct a graph $G(Q)$ with vertices $V_1, \ldots, V_n$ and lines indexed by $\Gamma$: the line $\gamma = (\ell, i, j) \in \Gamma$ connects the vertices $V_i$ and $V_j$ (we will take $i_\gamma = j$ and $f_\gamma = i$ [see Definition A.1], but actually this orientation is arbitrary). Then we have a 3rd equivalent condition:

(Q 2)″ The graph $G(Q)$ is connected.

The equivalence of (Q 2), (Q 2)′, and (Q 2)″ follows from the results of Appendix A. Suppose we assign to each line $\gamma = (\ell, i, j) \in G(Q)$ the inverse Feynman parameter

$$x_\gamma = \frac{Q_{ij}^{(\ell)}}{\alpha_\ell} ;$$

note that (Q 1) implies $x_\gamma > 0$. In Definition A.8 we defined a matrix $A[G]$ for any graph G, depending on inverse parameters associated with the lines of G. Then an easy calculation gives

$$A[G(Q)](x_\gamma) = A , \tag{2.25}$$

where A is given in (2.24). The equivalence of (Q 2), (Q 2)′, and (Q 2)″ now follows from Corollary A.10.

We note that if $e_i^{(\ell)}$ is the incidence matrix of some graph G, and $Q_{ij}^{(\ell)} = -e_i^{(\ell)} e_j^{(\ell)}$, then $G(Q) = G$ and (Q 2)″ reduces to the condition that G itself is connected. That is, our GFA is defined so that it interpolates between the Feynman amplitudes of connected graphs. However, (Q 1) and (Q 2) do in fact leave enough freedom in our choice of quadratic forms to enable us to interpolate between all graphs in the class discussed in the third paragraph of Section 1. We will return to this point in Chapter IV.

From now on, then, we will assume that Q satisfies (Q 1) and (Q 2). The only delicate point involved in discussing the limit (2.23) is the behavior of the quantities $[\det \mathcal{Q}_E]$ and $\mathcal{Q}_E^{-1}$ which appear in (2.17). We prove a series of lemmas about this behavior.

*Lemma 2.12:* Let A be defined by (2.24), and $\mathcal{Q} = \mathcal{Q}_1 + i\mathcal{Q}_2$ by (2.16). Then for $1 \leq i, j \leq n$,

(a). $\det \mathcal{Q}_{1E} = -n^4 [A(\begin{smallmatrix} i \\ i \end{smallmatrix})]^4$ .

(b). For any $\underline{p}, \underline{p}' \in E$, $\underline{p} \circ \mathcal{Q}_{1E}^{-1} \circ \underline{p}' = \dfrac{1}{A(\begin{smallmatrix} i \\ i \end{smallmatrix})} \Sigma_{r,s \neq j}\, p_r \cdot A(\begin{smallmatrix} j\, r \\ j\, s \end{smallmatrix}) \cdot p_s$ .

*Proof:* Since $\mathcal{Q}_1 = g \otimes A$, we must have

$$\det \mathcal{Q}_{1E} = -[\det A_E]^4 .$$

To calculate $\det A_E$ we must express $A_E$ as a transformation in E. Let $u^1, \ldots, u^n$ be standard basis vectors for $R^n$, so that $w \in R^n$ implies $w = \Sigma\, w_i u^i$. As a basis in E we may choose $v^i = u^i - u^n$ $(i = 1, \ldots, n-1)$. Then if $A_{ij}$ is the matrix of A in the original basis $[(Aw)_i = \Sigma_i A_{ij} w_j]$, the matrix of $A_E$ in this new basis is $(A_E)_{ij} = A_{ij} - A_{in}$ $(i, j = 1, \ldots, n-1)$. Thus

$$\det A_E = \det\{A_{ij} - A_{in} \mid i, j = 1, \ldots, n-1\} \,. \tag{2.26}$$

Now in the matrix on the right hand side of (2.26) add column 2 through (n−1) to column 1, then subtract 1/n times the new column 1 from each of the remaining columns. Using $\Sigma_{j=1}^{n} A_{ij} = 0$, this gives immediately

$$\det A_E = n\, A\binom{n}{1} \,.$$

But it is well known that all minors of A are equal (because A is translation invariant and symmetric), hence $\det A_E = n\, A\binom{i}{i}$ for any i.

(b). Since $\mathcal{Q}_1 = g \otimes A$, we have $\mathcal{Q}_{1E}^{-1} = g \otimes A_E^{-1}$. Now if $B_{ij}$ is a matrix on E written in the basis $\{v^i\}$ discussed above, and $w = \Sigma\, w_j u^j$, $w' = \Sigma\, w'_j u^j$ are any vectors in E, then $w' \circ B \circ w = \Sigma_{i,j=1}^{n-1} (w'_i - w'_n) B_{ij} w_j$ (to see this, note that $\{v_i = u^i - \frac{1}{n} \Sigma_1^n u^j \mid i = 1, \ldots, n-1\}$ is the dual basis to $v^i$ in the inner product 0 on E). We have already found the matrix of $A_E$ in the basis $\{v^i\}$; thus, the usual formula for matrix inversion gives

$$w' \circ A_E^{-1} \circ w = \frac{1}{\det A_E} \det C \,,$$

where

$$C = \begin{bmatrix} 0 & w'_1 - w'_n & & w'_{n-1} - w'_n \\ w_1 & A_{11} - A_{1n} & & A_{1,n-1} - A_{1,n} \\ \vdots & \vdots & \cdots & \vdots \\ w_{n-1} & A_{n-1,1} - A_{n-1,n} & & A_{n-1,n-1} - A_{n-1,n} \end{bmatrix}$$

We number the rows and columns of C 0 through (n−1), then manipulate C just as we did the matrix in (2.26). This yields $\det C = n\, D\binom{n}{1}$, where

$$D = \begin{bmatrix} 0 & w'_1 & \cdots & w'_n \\ w_1 & A_{11} & \cdots & A_{1,n} \\ \vdots & \vdots & & \vdots \\ w_n & A_{n,1} & \cdots & A_{n,n} \end{bmatrix} .$$

But again all minors $D\binom{k}{\ell}$ of D with $k, \ell > 0$ are equal so

$$w' \cdot A_E^{-1} \cdot w = \frac{1}{A\binom{i}{i}}\, D\binom{j}{j}$$

for any i,j. This yields part (b) of the Lemma.

*Lemma 2.13:* The matrix $A_E^{-1}$ is uniformly bounded in any region of the form

$$0 \le a_\ell \le M \quad (\text{all } \ell \in \mathfrak{L}). \tag{2.27}$$

*Proof:* From Lemma 2.12, it suffices to show that for r, s $\neq$ i, $A(\substack{i\,r\\i\,s})/A(\substack{i\\i})$ is uniformly bounded in (2.27). From (2.25) and Lemmas A.10, A.11, and A.13,

$$\frac{A(\substack{i\,r\\i\,s})}{A(\substack{i\\i})} = \frac{\Sigma_{T_2}\{\prod_{\gamma\in T_2} Q_{ij}^{(\ell)} \prod_{\gamma\notin T_2} a_\ell\}}{\Sigma_T\{\prod_{\gamma\in T} Q_{ij}^{(\ell)} \prod_{\gamma\notin T} a_\ell\}}, \tag{2.28}$$

where $\gamma = (\ell, i, j)$, and the sum in the numerator is over certain 2-trees of G(Q), the sum in the denominator over all trees of G(Q).

We study (2.28) in the region

$$0 \le a_{\ell_1} \le a_{\ell_2} \le \cdots \le a_{\ell_L} \le M, \tag{2.29}$$

and introduce new variables $t_1, \ldots, t_L$ defined by

$$a_{\ell_k} = t_L\, t_{L-1}, \ldots, t_k, \tag{2.30}$$

so that (2.29) becomes

$$\begin{aligned} &0 \le t_L \le M, \\ &0 \le t_\ell \le 1, \quad \text{if } 1 \le \ell < L. \end{aligned} \tag{2.31}$$

Let $G_\ell$ be the subgraph of G(Q) consisting of all lines $\gamma = (\ell', i, j)$ with $\ell' \le \ell$, and their vertices; let $N_\ell$ be the number of loops of $G_\ell$ (Definition A.1). If T is any tree in G(Q), and $T^*$ its complement, then $T^*$ must contain at least $N_\ell$ lines from $G_\ell$. Thus every term in the denominator of (2.28) contains a factor $t_1^{N_1} t_2^{N_2} \ldots t_L^{N_L}$; similarly, every term in the numerator contains this factor. On the other hand, we can find some tree $T_0$ such that $T_0{}^*$ contains exactly $N_1$ lines of $G_1$, $N_2$ lines of $G_2$, etc; thus

$$\prod_{\gamma\in T_0} a_\ell = t_1^{N_1} t_2^{N_2} \ldots t_L^{N_L}. \tag{2.32}$$

Thus if we cancel the factor (2.32) in the numerator and denominator of (2.28), the resulting denominator cannot vanish in the region (2.30); this proves the lemma.

*Remark 2.14:* Techniques similar to that used in the proof of Lemma 2.13, i.e., the change of variables (2.30), are quite important in discussing the analytic properties of the GFA in $\underline{\lambda}$. See Chapters III and IV for details.

We have already given the restrictions on Q under which we will prove the existence of the limit (2.23). We also mentioned a restriction on $\underline{\mathcal{Q}}_2$; it is that the limit be taken *nontangentially.* Thus let $K'$ be a convex compact set of real translation invariant quadratic forms positive definite on $E$, let $K$ be the cone on $K'$ with zero as vertex $(K = \{\mathcal{P} = t\mathcal{P}' | \mathcal{P}' \in K', 0 \leq t < \infty\})$, and let $\underline{K} = \{\underline{\mathcal{Q}}_2 | \mathcal{Q}_2^{(\ell)} \in K, \text{ for all } \ell\}$. Then we will prove the existence of (2.23) under the assumption $\underline{\mathcal{Q}}_2 \in \underline{K}$. We need one more lemma, and a corollary.

*Lemma 2.15:* Let $\mathcal{B} = \mathcal{B}_1 + i\mathcal{B}_2$ be a translation invariant quadratic form on $E$, with $\mathcal{B}_2 \in K$. Then if $\mathcal{B}_{1_E}$ is non singular,

(a) $|\det \mathcal{B}| > |\det \mathcal{B}_1|$,

(b) There is a constant $M$ depending only on $K$ such that

$$\|\mathcal{B}_E^{-1}\| < M \|\mathcal{B}_{1_E}^{-1}\| \ ;$$

(here $\| \ \|$ is the usual operator norm).

*Proof:* We work exclusively with matrices on $E$, and therefore drop $E$ as a subscript. There is a real unimodular matrix $\mathcal{C}$ such that $\mathcal{D} = \mathcal{C}^T\mathcal{B}\mathcal{C}$ is diagonal; moreover, $\|\mathcal{C}\|$ and $\|\mathcal{C}^{-1}\|$ are uniformly bounded for $\mathcal{B}_2 \in K$. [To see this, write $\mathcal{C} = \mathcal{C}_1\mathcal{C}_2\mathcal{C}_3$, where

(i) $\mathcal{C}_1$ is orthogonal and $\mathcal{C}_1^T\mathcal{B}_2\mathcal{C}$ is diagonal;

(ii) $\mathcal{C}_2$ is diagonal and unimodular, and $\mathcal{C}_2\mathcal{C}_1^T\mathcal{B}_2\mathcal{C}_1\mathcal{C}_2$ is a (positive) multiple of the identity;

(iii) $\mathcal{C}_3$ is orthogonal and $\mathcal{C}_3^T\mathcal{C}_2\mathcal{C}_1^T\mathcal{B}_1\mathcal{C}_1\mathcal{C}_2\mathcal{C}_3$ is diagonal.

Note $\mathcal{C}_1$ and $\mathcal{C}_3$ have norm 1, and the norms of $\mathcal{C}_2$ and $\mathcal{C}_2^{-1}$ depend only on $\mathcal{B}_2$, hence only on $K$.] Since we also have $\mathcal{D}_1 = \mathcal{C}^T\mathcal{B}_1\mathcal{C}$ and trivially

$$|\det \mathcal{D}| > |\det \mathcal{D}_1| \ ,$$

part (a) follows from the unimodularity of $\mathcal{C}$. Similarly, it is clear that

$$\|\mathcal{D}^{-1}\|_2 < \|\mathcal{D}_1^{-1}\|_2$$

where $\| \ \|_2$ is the Hilbert-Schmidt norm; using $\mathcal{B}^{-1} = \mathcal{C}\mathcal{D}^{-1}\mathcal{C}^T$, $\mathcal{B}_1^{-1} = \mathcal{C}\mathcal{D}_1^{-1}\mathcal{C}^T$ and the equivalence of the Hilbert-Schmidt and usual norms gives (b).

*Corollary 2.16:* Suppose $Q$ satisfies (Q 1), (Q 2), $\underline{\mathcal{Q}}_2 \in \underline{K}$, and $0 < a_\ell \leq M$ for all $\ell$. Then $\mathcal{A}_E^{-1}$ and $[\det \mathcal{A}_E]^{-1}$ are uniformly bounded and approach $\mathcal{A}_{1E}^{-1}$ and $[\det \mathcal{A}_{1E}]^{-1}$ as $\underline{\mathcal{Q}}_2 \to 0$.

*Proof:* Follows immediately from Lemmas 2.12, 2.13, and 2.15.

We are now prepared to prove the main theorem of this section.

*Theorem 2.17:* Let $Q$ satisfy (Q 1) and (Q 2), and suppose $\operatorname{Re}\lambda_\ell > (r_\ell + 2)$ $(\ell \in \mathcal{L})$. Then the limit

(2.33) $$\mathcal{T}(\underset{\sim}{\lambda}, \underset{\sim}{Q}, \underset{\sim}{e}) = \lim_{\underset{\sim}{\mathcal{Q}}_2 \to 0;\ \underset{\sim}{\mathcal{Q}}_2 \in \underset{\sim}{K}} \mathcal{T}(\underset{\sim}{\lambda}, \underset{\sim}{\mathcal{Q}}, \underset{\sim}{e})$$

exists [in $\mathcal{S}'(R^{4n})$] and is analytic in $\underset{\sim}{\lambda}$. We have the formula

$$\tilde{\mathcal{T}}(\underset{\sim}{\lambda}, \underset{\sim}{Q}, \underset{\sim}{e})(\underset{\sim}{p}) = \lim_{\epsilon \to 0} i\, b(\underset{\sim}{\lambda})\, \delta(\Sigma p_i) \int_0^\infty \cdots \int_0^\infty \prod_{\mathfrak{L}} \alpha_\ell^{\lambda_\ell - 3} d\alpha_\ell \times$$

$$\tau\left[\prod_{\mathfrak{L}} Z_\ell(Y^{(\ell)})\right] [A(\tfrac{i}{i})]^{-2} \exp i\left\{ \underset{\sim}{p} \cdot A_E^{-1} \cdot \underset{\sim}{p} - \sum_{\mathfrak{L}} \alpha_\ell (m_\ell^2 - i\epsilon)\right\} .$$

Here $b(\underset{\sim}{\lambda})$ is given in (2.18), and

(2.35 a) $$Y^{(\ell)} = -\frac{1}{2} e^{(\ell)} \cdot \frac{Q^{(\ell)}}{\alpha_\ell} \cdot A_E^{-1} \cdot \underset{\sim}{p}$$

(2.35 b) $$\overline{Y^{(\ell)} Y^{(\ell')}} = -\frac{i}{8} e^{(\ell)} \cdot \left[\frac{Q^{(\ell)}}{\alpha_\ell} \cdot A_E^{-1} \cdot \frac{Q^{(\ell')}}{\alpha_{\ell'}} + \delta_{\ell\ell'} \frac{Q}{\alpha_\ell}\right] \cdot e^{(\ell')} g$$

*Proof:* From (2.17) and Lemma 2.10,

(2.36) $$\tilde{\mathcal{T}}(\underset{\sim}{\lambda}, \underset{\sim}{\mathcal{Q}}, \underset{\sim}{e})(\underset{\sim}{p}) = b(\underset{\sim}{\lambda})\, \delta_E \int_0^\infty \cdots \int_0^\infty \prod_{\mathfrak{L}} [\alpha_\ell^{\lambda_\ell - 3} d\alpha_\ell]\, \tau \left[\prod_{\mathfrak{L}} Z_\ell(X^{(\ell)})\right]$$
$$[\det \mathcal{a}_E]^{-1/2} \exp i[\underset{\sim}{p} \circ G \circ \mathcal{a}_E^{-1} \circ G \circ \underset{\sim}{p} \circ - \Sigma\, \alpha_\ell m_\ell^2] .$$

We expand the $\tau$-product in (2.36) and show that each term approaches the corresponding term of the expansion of the $\tau$-product in (2.34). A typical term from (2.36) has the form

(2.37) $$X^{(\ell_1)} \cdots X^{(\ell_m)}\, \overline{X^{(\ell_{m+1})} X^{(\ell_{m+2})}} \cdots \overline{X^{(\ell_{m+2m'-1})} X^{(\ell_{m+2m'})}}$$

where each $\ell \in \mathfrak{L}$ may occur at most $\underset{\sim}{r_\ell}$ times among $\ell_1, \ldots, \ell_{m+2m'}$. Note that $X^{(\ell)}$ is homogeneous of degree zero in $\alpha$, $\overline{X^{(\ell)} X^{(\ell')}}$ homogeneous of degree $-1$ (equation 2.21).

We now make a change of variables in (2.36):

$$t = \sum_{\mathfrak{L}} \alpha_\ell ;$$

$$\alpha_\ell = \beta_\ell t , \quad \sum_{\mathfrak{L}} \beta_\ell = 1 .$$

The Jacobian of this change is $t^{L-1}$. Using the homogeneity of $X^{(\ell)}$, $\overline{X^{(\ell)} X^{(\ell')}}$, $\mathcal{a}_E^{-1}$, and $\det \mathcal{a}_E$ in $\alpha$, we can do the $t$-integral explicitly by Lemma 2.8 (a). Thus the term in (2.36) corresponding to (2.37) is equal to

$$(2.38)\qquad \begin{aligned} & e^{\nu\pi i/2}\Gamma(\nu)\, b(\underline{\lambda})\, \delta_E \int_{\Sigma\beta_\ell = 1} \left[\prod_{\mathcal{L}} d\beta_\ell \beta_\ell^{\lambda_\ell - 3}\right] [\det \mathcal{A}_E']^{-1/2} \\ & \prod_{i=1}^{m} X^{(\ell_i)'} \prod_{i=1}^{m'} \overline{X^{(\ell_{m+2i-1})'}\, X^{(\ell_{m+2i})'}}\, [\underline{p} \circ G \circ \mathcal{A}_E^{-1} \circ G \circ \underline{p} - \Sigma\, \beta_\ell m_\ell^2]^{-\nu}, \end{aligned}$$

where $\nu = \Sigma_{\mathcal{L}}\, \lambda_\ell - 2L - m' + (2n-2)$, and the primes on $\mathcal{A}_E$ and $X^{(\ell)}$ indicate that $\alpha_\ell$ is to be replaced by $\beta_\ell$ throughout.

Let us smear (2.38) with a test function f and then take the limit $\underline{\mathcal{Q}}_2 \to 0$, $\underline{\mathcal{Q}}_2 \in \underline{K}$. We will prove in a moment that

$$(2.39)\qquad [\det \mathcal{A}_E']^{-1/2} \to i[n\, A(\tfrac{i}{i})]^{-2}$$

(this almost follows from Lemma 2.12; the only question is the sign). Similarly

$$\mathcal{A}_E^{-1} \to g \otimes A_E^{-1},$$

$$X^{(\ell)} \to Y^{(\ell)},$$

$$\overline{X^{(\ell)} X^{(\ell')}} \to \overline{Y^{(\ell)} Y^{(\ell')}}.$$

This means that the $\beta$-integrand of (2.38) smeared with f approaches the corresponding $\beta$-integrand for (2.34) pointwise, using Theorem B.8. Moreover, Corollary (2.16) implies that the $\beta$-integrand is uniformly bounded by an integrable function; the only question here is the behavior of $\overline{X^{(\ell)} X^{(\ell')}}$, which contains a factor $\frac{1}{\beta_\ell}\ \frac{1}{\beta_{\ell'}}$. However, writing $\beta_\ell^{\lambda_\ell - 3} = \beta_\ell^{\lambda_\ell - r_\ell - 3}\, \beta_\ell^{r_\ell}$, we see that the factor $\beta_\ell^{\lambda_\ell - r_\ell - 3}$ is integrable while the remaining factor $\beta_\ell^{r_\ell}$ suffices to make (2.37) uniformly bounded. Thus by the Lebesgue dominated convergence theorem, (2.38) is equal to

$$\begin{aligned} & i\, e^{\nu\pi i/2}\Gamma(\nu)\, b(\underline{\lambda})\, \delta(\Sigma\, p_i) \int_{\Sigma\beta_\ell = 1} [\Pi\, d\beta_\ell \beta_\ell^{\lambda_\ell - 3}]\, [A(\tfrac{i}{i})']^{-2} \\ & \prod_{1}^{m} Y^{(\ell_i)'} \prod_{1}^{m'} \overline{Y^{(\ell_{m+2i-1})'}\, Y^{(\ell_{m+2i})'}}\, [\underline{p} \cdot A_E^{-1'} \cdot \underline{p} - \Sigma\, \beta_\ell m_\ell^2]^{-\nu}. \end{aligned}$$

Another application of Theorem B.8 (to insert an $i\epsilon\, \Sigma\, \beta_\ell$ in 2.40) and Lemma 2.8(a) completes the proof. Theorem B.8 also implies that (2.34) is analytic in $\underline{\lambda}$.

There remains only to check the sign in (2.39). The sign of $[\det \mathcal{A}_E]^{-1/2}$ is determined by the requirement that this quantity be positive when $i\,\mathcal{A}_E$ is real and positive definite. Let B be translation invariant with B positive definite on E, and let $\mathcal{B}(\theta) = g(\theta) \otimes B$, with

$$g(\theta) = \begin{bmatrix} -i e^{i\theta} & & & \\ & -i e^{-i\theta} & & 0 \\ & & -i e^{-i\theta} & \\ 0 & & & -i e^{-i\theta} \end{bmatrix}$$

$\mathcal{B}(\theta)$ has the following properties:

(i) $i\mathcal{B}(0)$ is positive definite on $E$.

(ii) $\mathrm{Re}[i\mathcal{B}(\theta)]$ is positive definite for $0 \le \theta < \pi/2$.

(iii) $\mathcal{B}(\pi/2) = g \otimes B$.

Thus if we define $[\det \mathcal{B}_E(0)]^{1/2} > 0$ we have

$$[\det \mathcal{B}_E(0)]^{1/2} = e^{-i\theta}[\det B_E]^2$$

for all $\theta$, so that $[\det g \otimes B]^{1/2} = -i[\det B_E]^2$. This was derived for $B$ positive definite on $E$ but holds by analytic continuation for all $B$, in particular, for $B = A$. This proves (2.39), and completes the proof of Theorem 2.17.

*Remark 2.18:* Formula (2.34) is our expression for the GFA; it is as yet valid only for $\mathrm{Re}\,\lambda_\ell > r_\ell + 2$. However, we mentioned in Section 4 the possibility of a simplification in this expression. When $\epsilon_j^{(\ell)}$ is in fact the incidence matrix of some graph, and $Q_{rs}^{(\ell)} = -e_r^{(\ell)} e_s^{(\ell)}$, then (2.35) reduces to

(2.40 a)
$$Y^{(\ell)} = \frac{1}{2} \frac{e^{(\ell)}}{\alpha_\ell} \cdot A_E^{-1} \cdot \underline{p} \;,$$

(2.40 b)
$$\overline{Y^{(\ell)} Y^{(\ell')}} = -\frac{i}{2}\left[\frac{e^{(\ell)}}{\alpha_\ell} \cdot A_E^{-1} \cdot \frac{e^{(\ell')}}{\alpha_{\ell'}} - \frac{\delta_{\ell\ell'}}{\alpha_\ell}\right] g \,,$$

using $e^{(\ell)} \cdot Q^{(\ell)} = -2e^{(\ell)}$, $e^{(\ell)} \cdot e^{(\ell)} = 2$. If we now *define* quantities $\hat{Y}^{(\ell)}$ and $\overline{\hat{Y}^{(\ell)}\hat{Y}^{(\ell')}}$ by the right hand sides of (2.40), we may define a new GFA by

(2.42)
$$\tilde{\hat{\mathcal{T}}}(\underline{\lambda}, \underline{Q}, \underline{e}) = \lim_{\epsilon \to 0} i\, b(\lambda)\, \delta(\Sigma\, p_i) \int_0^\infty \cdots \int_0^\infty \prod_{\mathcal{L}} \alpha_\ell^{\lambda_\ell - 3}\, d\alpha_\ell$$
$$\tau[\prod_{\mathcal{L}} Z_\ell(\hat{Y}^{(\ell)})]\; A(\tfrac{i}{i})^{-2} \exp i\{\underline{p} \cdot A_E^{-1} \cdot \underline{p} - \sum_{\mathcal{L}} \alpha_\ell(m_\ell^2 - i\epsilon)\} \,.$$

Now $\hat{\mathcal{T}}$ works just as well as $\mathcal{T}$ as a GFA, and is slightly simpler in form. Most of our future work will be independent of which GFA we use; when we wish to specialize to one or the other, we will say so explicitly. It should be emphasized that they are equivalent when $\underline{e}$ and $\underline{Q}$ arise from a graph.

## CHAPTER III

### *Analytic Renormalization*

Section 1. INTRODUCTION.

In this chapter we discuss the GFA corresponding to an actual connected Feynman graph $G_0$. That is, we discuss (2.34) [or equivalently (2.42)], taking $\underline{e}$ to be the incidence matrix of the graph $G_0$ and $Q_{ij}^{(\ell)} = -e_i^{(\ell)} e_j^{(\ell)}$. Our main interest will be in the dependence of the GFA on the complex variables $\lambda_1, \ldots, \lambda_L$. Much of this chapter is taken from the author's paper [33]; in addition, a simpler derivation of (2.34) (or an equivalent formula) is given there.

Now (2.34) has been established only for

$$\text{Re}\, \lambda_\ell > r_\ell + 2, \qquad (\ell = 1, \ldots, L) \tag{3.1}$$

while the point of physical significance is $\lambda_\ell = 1$ $(\ell = 1, \ldots, L)$. (At $\lambda_\ell = 1$ the GFA becomes formally equal to the Feynman amplitude of the graph.) Our first task will be to extend the definition of the GFA to all $\underline{\lambda} \in \mathbb{C}^L$. This is done by a process of analytic continuation: the GFA is analytic in $\underline{\lambda}$ for $\underline{\lambda}$ satisfying (3.1), hence any extension is essentially uniquely determined. We will show that such an extension does exist, and that the resulting amplitude is meromorphic on $\mathbb{C}^L$; finally, we will give explicitly the possible singularities of the amplitude and discuss their relation to the structure of $G_0$.

We would now like to use this analytically continued GFA to define the Feynman amplitude of the graph. However, as might be expected, the point $\lambda_\ell = 1$ is a singularity of the GFA precisely when the integral defining the original Feynman amplitude of $G_0$ is divergent. Nevertheless, it is possible to define an amplitude for the point $\lambda_\ell = 1$ by "subtracting off" the singularities of the GFA in an appropriate way. If the GFA $\mathcal{T}$ depended on only one complex variable $z$ and had a pole at $z = 1$, we could "subtract off" the singularity by discarding all terms in the Laurent series for $\mathcal{T}$ at $\lambda = 1$ which had negative exponents; this would give us as a finite part for $\mathcal{T}$ simply the constant term of the Laurent series. This method of defining divergent quantities originated with M. Riesz [28] and has been used in various contexts by many authors [see, e.g., 4, 10, 15, 21, 22, 26, 30]. In this chapter we show how to generalize this procedure to the several complex variables $\lambda_1, \ldots, \lambda_L$. Our criterion for the procedure is that the "finite part" of the GFA which it produces belong to the class of renormalized amplitudes of $G_0$ defined in Definition 1.19. We prove that our procedure satisfies this condition.

Section 2. ANALYTIC PROPERTIES.

In this section we will work exclusively with the GFA for the graph $G_0$; we write this GFA simply as $\mathcal{T}(\underline{\lambda})$. From (2.42) we have

$$\mathcal{T}(\underline{\lambda}) = \lim_{\epsilon \to 0} \mathcal{T}_\epsilon(\underline{\lambda}) \; ,$$

(3.2)
$$\mathcal{T}_\epsilon(\underline{\lambda})(\underline{p}) = i\, b(\underline{\lambda})\, \delta\Big(\sum_{i=1}^{n} p_i\Big) \int_0^\infty \cdots \int_0^\infty \Big[\prod_{\mathcal{L}} \alpha_\ell^{\lambda_\ell - 3}\, d\alpha_\ell\Big] \times \tau \Big[\prod_{\mathcal{L}} Z_\ell(Y^{(\ell)})\Big] A(\tfrac{i}{i})^{-2} \exp i \Big\{ \underline{p}\cdot A_E^{-1}\cdot \underline{p} - \sum_{\mathcal{L}} \alpha_\ell (m_\ell^2 - i\epsilon) \Big\}$$

where $A = A(G_0)(\alpha_1, \ldots, \alpha_L)$ (Definition A.8) and, if $e_i^{(\ell)}$ is the incidence matrix of $G_0$,

(3.3 a)
$$Y^{(\ell)} = \frac{e^{(\ell)}}{\alpha_\ell} \cdot A_E^{-1} \cdot \underline{p} \quad ,$$

(3.3 b)
$$\overline{Y^{(\ell)}\, Y^{(\ell')}} = -\frac{i}{2}\Big[\frac{e^{(\ell)}}{\alpha_\ell} \cdot A_E^{-1} \cdot \frac{e^{(\ell')}}{\alpha_{\ell'}} - \frac{\delta_{\ell\ell'}}{\alpha_\ell}\Big] g \; ,$$

and

(3.4)
$$b(\underline{\lambda}) = \frac{2^{2n-4L}\, \pi^{2n-2L-2}\, (-i)^{2n-2+L}}{\prod_{\mathcal{L}} \Gamma(\lambda_\ell)} \; .$$

Before discussing the analytic properties of $\mathcal{T}(\underline{\lambda})$, we give some results on the structure of some subgraphs of $G_0$; this structure will be related to the singularities of $\mathcal{T}(\underline{\lambda})$. The basic terminology is given in Definitions A.1 to A.3.

*Definition 3.1:* Let $G_0$ be a connected graph, as above. A *singularity family* (s-family) for $G_0$ is a maximal collection $\mathfrak{S}$ of (non-empty) irreducible subgraphs of $G_0$ satisfying

(S1) if $G, G' \in \mathfrak{S}$ then either $G \subset G'$, $G' \subset G$, or $\mathcal{L}(G) \cap \mathcal{L}(G') = \emptyset$ ;

(S2) if $G_1, \ldots, G_k \in \mathfrak{S}$, and $\mathcal{L}(G_i) \cap \mathcal{L}(G_j) = \emptyset$ for any i, j, then $G = \cup_{i=1}^{k} G_i$ is not irreducible.

A *labeled s-family* is a pair $(\mathfrak{S}, \sigma)$, where $\mathfrak{S}$ is an s-family and $\sigma$ is a mapping $\sigma : \mathfrak{S} \to \mathcal{L}(G_0)$ satisfying

(S3) $\sigma(G) \in \mathcal{L}(G)$ ;

(S4) if $G' \in \mathfrak{S}$ is a proper subgraph of $G \in \mathfrak{S}$, then $\sigma(G) \notin \mathcal{L}(G')$.

Finally, if $(\mathfrak{S}, \sigma)$ is a labeled s-family, we define $\mathcal{D}(\mathfrak{S}, \sigma)$ to be the subset of $\alpha$-space given by

$$\mathcal{D}(\mathfrak{S}, \sigma) = \{(\alpha_1, \ldots, \alpha_L) \mid \alpha_\ell \geq 0 \text{ for all } \ell,\ \alpha_\ell \leq \alpha_{\sigma(G)} \text{ for } \ell \in G \in \mathfrak{S}\}.$$

*Lemma 3.2:* Let $\mathfrak{S}$ be an s-family. For any $G \in \mathfrak{S}$, let $\mathfrak{S}(G)$ consist of all elements of $\mathfrak{S}$ which are proper subgraphs of G. Then:

(a) For any $G \in \mathfrak{S}$, $N(G) = \#[\mathfrak{S}(G)] + 1$.

(b) $\#(\mathfrak{S}) = N\ [= N(G_0)]$ .

(c) $\mathfrak{S}$ may be labeled, i.e., there is a $\sigma: \mathfrak{S} \to \mathfrak{L}(G)$ such that $(\mathfrak{S}, \sigma)$ is a labeled s-family.

(d). If $(\mathfrak{S}, \sigma)$ is a labeled s-family, then $\mathfrak{L}(G_0) - \sigma(\mathfrak{S})$ is a tree in $G_0$ (see Definition A.5).

*Proof:* (a). Take $G \in \mathfrak{S}$, and let $\ell \in \mathfrak{L}(G)$ be a line not contained in $\mathfrak{L}(G')$ for any $G' \in \mathfrak{S}(G)$ [such a line exists by (S2)]. Let $H_1, \ldots, H_k$ be the irreducible components of $G_1$, the graph formed from G by deleting $\ell$. The maximality condition on $\mathfrak{S}$ then implies $H_1, \ldots, H_k \in \mathfrak{S}$, since $\mathfrak{S} \cup \{H_i\}$ satisfies (S1) and (S2).

Now we prove (a) by induction on $\#[\mathfrak{S}(G)]$. If $\#[\mathfrak{S}(G)] = 0$, we see by the above that $G_1$ has no irreducible components, whence $N(G_1) = 0$, $N(G) = 1$. In general, since every $G' \in \mathfrak{S}(G)$ must be contained in some $H_i$,

$$
\begin{aligned}
N(G) &= N(G_1) + 1 \\
&= \sum_{i=1}^{k} N(H_i) + 1 \qquad \text{(see Remark A.4)} \\
&= \sum_{i=1}^{k} \{\#[\mathfrak{S}(H_i)] + 1\} + 1 \quad \text{(induction assumption)} \\
&= \#[\mathfrak{S}(G)] + 1 \ .
\end{aligned}
$$

(b). Let $H_1, \ldots, H_k$ be the irreducible components of $G_0$. Again, the maximality condition implies $H_i \in \mathfrak{S}$. Then

$$
\begin{aligned}
N = N(G_0) &= \Sigma\, N(H_i) \\
&= \Sigma\, \{\#[\mathfrak{S}(H_i)] + 1\} \\
&= \#\, \mathfrak{S}(H) \ ,
\end{aligned}
$$

since every $G' \in \mathfrak{S}$ must be contained in some $H_i$.

(c). As pointed out in (a), there exists for each $G \in \mathfrak{S}$ an $\ell \in \mathfrak{L}(G)$ with $\ell \notin \mathfrak{L}(G')$ for any $G' \in \mathfrak{S}(G)$. Take $\sigma(G) = \ell$.

(d). Let $(\mathfrak{S}, \sigma)$ be a labeled s-family, let $T = \mathfrak{L}(G_0) - \sigma(\mathfrak{S})$, and let H be the graph formed by T and all vertices of the lines in T. Now if $G, G' \in \mathfrak{S}$ with $G \neq G'$, we must have $\sigma(G) \neq \sigma(G')$; then part (b) implies that $\#(T) = L - N = n-1$. It therefore suffices to prove that every vertex of $G_0$ lies in H.

Suppose that V is a vertex of $G_0$ with $V \notin H$. Let $\mathfrak{L}' = \{\ell \in \mathfrak{L} \mid V_{i_\ell} = V \text{ or } V_{f_\ell} = V\}$. Then $\mathfrak{L}' \subset \sigma(\mathfrak{S})$; let $\mathfrak{S}' = \sigma^{-1}(\mathfrak{L}')$ and let G be a minimal element of $\mathfrak{S}'$ [such elements exist by (S1)]. Because G is irreducible, $\#[\mathfrak{L}' \cap \mathfrak{L}(G)] \geq 2$; so there must exist an $\ell' = \sigma(G') \in [\mathfrak{L}' \cap \mathfrak{L}(G)]$ with $G' \neq G$. But then (S1) and the minimality of G imply $G \subset G'$, and this contradicts (S4).

*Lemma 3.3:* Let $G_0$ be a connected graph. Then:

(a).

$$\cup\, \mathfrak{D}(\mathfrak{S}, \sigma) = \{\underline{a} \mid a_\ell \geq 0, \text{ all } \mathfrak{L}\} \ , \tag{3.5}$$

where the union is over all labeled s-families of $G_0$.

(b). If $(\mathfrak{S},\sigma)$ and $(\mathfrak{S}',\sigma')$ are two different labeled s-families of $G_0$, then $\mathfrak{D}(\mathfrak{S},\sigma) \cap \mathfrak{D}(\mathfrak{S}',\sigma')$ has Lebesgue measure zero (it is a subset of $R^L$).

*Proof:* (a). Let $\underline{a} = (a_1, \ldots, a_L)$ be a point satisfying $a_\ell \geq 0$, and choose $\ell_1, \ldots, \ell_L$ so that $a_{\ell_1} \leq a_{\ell_2} \leq \cdots \leq a_{\ell_L}$. Let $G_k$ be the graph consisting of lines $\ell_1, \ldots, \ell_k$ and all their endpoints, define $\mathfrak{S} = \{G \mid G$ is an irreducible component of $G_k$ for some $k\}$, and define $\sigma: \mathfrak{S} \to \mathfrak{L}$ by $\sigma(G) = \ell_{j(G)}$, where $j(G) = \max\{i \mid \ell_i \in G\}$. We will show that $(\mathfrak{S},\sigma)$ is a labeled s-family with $\underline{a} \in \mathfrak{D}(\mathfrak{S},\sigma)$.

Suppose first $G, G' \in \mathfrak{S}$ are irreducible components of $G_k$ and $G_k'$, respectively, with $k' \geq k$ and hence $G_k \subset G_k'$. Let $G''$ be the irreducible component of $G_k'$ which contains $G$; then either $G'' = G'$, in which case $G \subset G'$, or $G'' \neq G'$, in which case $\mathfrak{L}(G) \cap \mathfrak{L}(G') = \emptyset$. Thus $\mathfrak{S}$ satisfies (S1).

Now suppose $H_1, \ldots, H_k \in \mathfrak{S}$ are such that $\cup_{i=1}^k H_i$ is irreducible. Let $H_i$ be an irreducible component of $G_{j_i}$, with $j_1 \leq j_2 \leq \cdots \leq j_k$. Then $\cup_{i=1}^k H_i \subset G_{j_k}$ and hence $\cup_{i=1}^k H_i = H_k$; this proves (S2). Finally, since $N(G_L) = N$, there exist $j_1 \leq \cdots \leq j_N$ such that

$$N(G_{j_i}) = N[G_{(j_i-1)}] + 1, \qquad (i = 1, \ldots, N).$$

This means that $G_{j_i}$ has some irreducible component not contained in $G_{(j_i-1)}$, hence $\#(\mathfrak{S}) \geq N$. Combined with (b) of Lemma 3.2 this implies $\mathfrak{S}$ is maximal and hence is an s-family.

Now take $G \in \mathfrak{S}$, and suppose $G' \in \mathfrak{S}(G)$ is an irreducible component of $G_k'$. We must show $\sigma(G) \notin \mathfrak{L}(G')$. But $\sigma(G) \in \mathfrak{L}(G')$ implies $j(G) \leq k'$, since $\sigma(G) = \ell_{j(G)}$, and by definition of $j(G)$ this implies $i \leq k'$ for each $\ell_i \in G$, i.e., $G \subset G_k'$. But then $G'$ an irreducible component of $G_k'$ implies $G = G'$; this contradiction proves that $\sigma$ is a labeling. The fact that $\underline{a} \in \mathfrak{D}(\mathfrak{S},\sigma)$ follows immediately from the definition of $\sigma$.

(b). Now let $(\mathfrak{S},\sigma)$ and $(\mathfrak{S}',\sigma')$ be different labeled s-families, and take $\underline{a} \in \mathfrak{D}(\mathfrak{S},\sigma) \cap \mathfrak{D}(\mathfrak{S}',\sigma')$. We will show that $a_\ell = a_{\ell'}$ for some $\ell \neq \ell'$; the set of such $\underline{a}$'s has Lebesgue measure zero. First note that if $\mathfrak{S} = \mathfrak{S}'$, then $\sigma(G) \neq \sigma'(G)$ for some $G \in \mathfrak{S}$, and hence $a_{\sigma(G)} = a_{\sigma'(G)}$ by the definition of $\mathfrak{D}(\mathfrak{S},\sigma)$.

Now suppose $\mathfrak{S} \neq \mathfrak{S}'$. For any $G \in \mathfrak{S}$, define $\mathfrak{S}_0(G)$ to be the set of maximal elements of $\mathfrak{S}(G)$; similarly define $\mathfrak{S}_0'(G)$ for $G \in \mathfrak{S}'$. Now all irreducible components of $G$ lie in both $\mathfrak{S}$ and $\mathfrak{S}'$ (by the maximality condition in Definition 3.1), so $\mathfrak{S} \neq \mathfrak{S}'$ implies $\mathfrak{S}_0(G) \neq \mathfrak{S}_0'(G)$ for some $G \in \mathfrak{S} \cap \mathfrak{S}'$. But $\mathfrak{S}(G)$ is precisely the set of irreducible components of the graph obtained by deleting $\sigma(G)$ from $G$; thus we must have $\sigma(G) \neq \sigma'(G)$. But as above, $a_{\sigma(G)} = a_{\sigma'(G)}$; this completes the proof.

After these preliminaries on s-families, we turn to the analytic continuation of the Feynman amplitude $\mathcal{J}(\underline{\lambda})$. Note that the following theorem explains the terminology "singularity family."

*Theorem 3.4:* The distribution $T_\epsilon(\underline{\lambda})$ [equation (3.2)] may be written

$$\mathcal{T}_\epsilon(\underline{\lambda}) = \sum_{\mathfrak{S}} \mathcal{T}_\epsilon(\underline{\lambda}, \mathfrak{S}), \tag{3.6}$$

the sum extending over all s-families $\mathfrak{S}$. Here $\mathcal{T}_\epsilon(\underline{\lambda}, \mathfrak{S})$ has the property that the function

$$\prod_{G \in \mathfrak{S}} \left\{ \Gamma\left( \sum_{\ell \in \mathfrak{L}(G)}' \lambda_\ell - L(G) - \left[\frac{\mu(G)}{2}\right] \right) \right\}^{-1} \mathcal{T}_\epsilon(\underline{\lambda}, \mathfrak{S}) \tag{3.7}$$

may be extended from the region (3.1) to an entire analytic function of $\underline{\lambda} \in \mathbb{C}^L$. Here $\mu(G)$ is the superficial divergence of G (Definition 1.4).

*Remark 3.5:* This theorem describes all possible singularities of $\mathcal{T}_\epsilon(\underline{\lambda}, \mathfrak{S})$ and hence of $\mathcal{T}_\epsilon(\underline{\lambda})$. Specifically, to each $G \in \mathfrak{S}$ there corresponds a sequence of simple poles of $\mathcal{T}_\epsilon(\underline{\lambda}, \mathfrak{S})$, occurring on hyperplanes (in $\mathbb{C}^L$) of the form

$$\sum_{\ell \in \mathfrak{L}(G)}' \lambda_\ell = L(G) + \left[\frac{\mu(G)}{2}\right] - m, \tag{3.8}$$

where $m = 0, -1, -2, \ldots$ . Note that (3.8) may be written in the form

$$\sum_{\ell \in \mathfrak{L}(G)}' (\lambda_\ell - 1) = \left[\frac{\mu(G)}{2}\right] - m .$$

Thus if the superficial divergence $\mu(G)$ is negative for all irreducible subgraphs G of $G_0$, $\mathcal{T}_\epsilon(\underline{\lambda})$ is analytic at the point $\lambda_\ell = 1$ of physical interest; this corresponds to complete convergence of all integrals in the Feynman amplitudes (see Chapter I, Section 2.B). More generally, however, we see that $\mathcal{T}_\epsilon$ may have poles on a great many hyperplanes passing through the point $\lambda_\ell = 1$.

*Proof of Theorem 3.4:* For any labeled s-family $(\mathfrak{S}, \sigma)$, we define

$$\begin{aligned} \mathcal{T}_\epsilon(\underline{\lambda}, \mathfrak{S}, \sigma) &= i\, b(\underline{\lambda})\, \delta\left(\sum_{i=1}^{n} p_i\right) \int_{\mathcal{P}(\mathfrak{S},\sigma)} d\underline{\alpha} \prod_{\mathfrak{L}} \alpha_\ell^{\lambda_\ell - 3} \\ &\times \tau\left[\prod_{\mathfrak{L}} Z_\ell\,(Y^{(\ell)})\right] A(\tfrac{i}{i})^{-2} \exp i \left\{ \underline{p} \cdot A_E^{-1} \cdot \underline{p} - \sum_{\mathfrak{L}} \alpha_\ell (m_\ell^2 - i\epsilon) \right\}, \end{aligned} \tag{3.9}$$

then set

$$\mathcal{T}_\epsilon(\underline{\lambda}, \mathfrak{S}) = \sum \mathcal{T}_\epsilon(\underline{\lambda}, \mathfrak{S}, \sigma) ,$$

where the sum is over all labelings $\sigma$ of the s-family $\mathfrak{S}$. Then (3.2) and Lemma 3.3 imply that the $\mathcal{T}_\epsilon(\underline{\lambda}, \mathfrak{S})$ satisfy (3.6). We will actually prove the analytic properties stated in the theorem separately for each $\mathcal{T}_\epsilon(\underline{\lambda}, \mathfrak{S}, \sigma)$ .

Thus we take a fixed labeled s-family $(\mathfrak{S}, \sigma)$. Let $T_0 = \mathfrak{L}(G_0) - \sigma(E)$ [see Lemma 3.2)], $\mathfrak{L}' = \cup_{G \in \mathfrak{S}} \mathfrak{L}(G)$, and for $\ell \in \mathfrak{L}'$, let $G(\ell)$ be the minimal element of $\mathfrak{S}$

which contains $\ell$. Then in the region $\mathcal{D}(\mathfrak{S},\sigma)$ we introduce new variables $\{\gamma_\ell, t_G, \beta_{\ell'} \mid \ell \in (\mathcal{L}-\mathcal{L}'),\ G \in \mathfrak{S},\ \ell' \in [\mathcal{L}'-\sigma(\mathfrak{S})]\}$ related to the $\alpha_\ell$'s by:

$$\alpha_\ell = \begin{cases} \prod_{G'\supset G,\, G'\in\mathfrak{S}} t_{G'}, & \text{if } \ell = \sigma(G),\ G \in \mathfrak{S}; \quad (3.10a) \\ \beta_\ell \prod_{G'\supset G(\ell),\, G'\in\mathfrak{S}} t_{G'}, & \text{if } \ell \in [\mathcal{L}'-\sigma(\mathfrak{S})]; \quad (3.10b) \\ \gamma_\ell, & \text{if } \ell \in (\mathcal{L}-\mathcal{L}'). \quad (3.10c) \end{cases}$$

For notational convenience we sometimes write $\beta_{\sigma(G)} = 1$, so that (3.10b) holds for $\ell = \sigma(G)$ also. Then if $H_1, \ldots, H_k$ are the irreducible components of $G_0$, the domain $\mathcal{D}(\mathfrak{S},\sigma)$ becomes (in the new variables) a domain $\mathcal{D}'(\mathfrak{S},\sigma)$ specified by

$$\begin{array}{ll} 0 \le \gamma_\ell < \infty & \ell \in (\mathcal{L}-\mathcal{L}')\,, \\ 0 \le t_{H_i} < \infty & i = 1,\ldots,k\,, \\ 0 \le t_G \le 1 & G \in \mathfrak{S},\ G \ne H_i\,, \\ 0 \le \beta_\ell \le 1 & \ell \in [\mathcal{L}'-\sigma(G)]\,. \end{array}$$

We now give two lemmas about the properties of the integrand of (3.9) in terms of the new variables.

*Lemma 3.6:* For any $1 \le i \le n$,

$$\alpha_1 \cdots \alpha_L\, A\binom{i}{i} = \left[\prod_{G\in\mathfrak{S}} t_G^{N(G)}\right] B(\underline{t},\underline{\beta})\,, \tag{3.11}$$

where $B(t,\beta)$ is a polynomial which is independent of $t_{H_i}$ $(i = 1,\ldots,k)$ and, more importantly, does not vanish in $\mathcal{D}'(\mathfrak{S},\sigma)$.

*Proof:* From Lemma A.9,

$$\alpha_1 \cdots \alpha_L A\binom{i}{i} = \sum_T \prod_{\ell \notin T} \alpha_\ell\,, \tag{3.12}$$

the sum taken over all trees $T$ of $G_0$. Note that (3.12) does not depend on any $\alpha_\ell$ with $\ell \notin \mathcal{L}'$, since any such $\ell$ is contained in every tree. Now since $G \in \mathfrak{S}$ has $N(G)$ loops, there must be at least $N(G)$ lines of $G$ not contained in $T$; since each $\alpha_\ell$ for $\ell \in G$ contains a factor $t_G$, we have the factorization (3.11). Since any $\ell \notin T$ must lie in some $H_i$, each term in the sum (3.12) contains $t_{H_i}$ *precisely* as $t_{H_i}^{N(H_i)}$, so $B$ is independent of $t_{H_i}$. Finally, the term in (3.10) with $T = T_0$ is

$$\prod_{G\in\mathfrak{S}} t_G^{N(G)}\,.$$

Thus in $\mathcal{D}'(\mathfrak{S},\sigma)$ $B$ is the sum of non-negative terms one of which (from the tree $T_0$) is 1; this shows $B(\underline{t},\underline{\beta}) \ne 0$ in $\mathcal{D}'(\mathfrak{S},\sigma)$.

*Lemma 3.7:* Let $\underline{p}$ be any vector in E (Definition 2.1). Then the following quantities are all $C^\infty$ on $\mathfrak{D}'(\mathfrak{S},\sigma)$ (in the variables $\underline{\gamma}, \underline{\beta}; \underline{t}$ ):

(a) $\underline{p} \cdot A_E^{-1} \cdot \underline{p}$ ;

(b) $Y^{(\ell)}$ (any $\ell \in \mathfrak{L}$);

(c) $\overline{Y^{(\ell)} Y^{(\ell')}} [\Pi_{\ell,\ell' \in G} t_G]$ (any $\ell \neq \ell' \in \mathfrak{L}$);

(d) $\overline{Y^{(\ell)} Y^{(\ell)}} [\Pi_{\ell \in G} t_G]$ (any $\ell \in \mathfrak{L}$) .

(Recall that a function is $C^\infty$ on a closed set if it is $C^\infty$ on some open neighborhood.)

*Proof:* The proof in each case consists of displaying the quantity in question as the ratio of two polynomials, the denominator being $B(\underline{t}, \underline{\beta})$ [Lemma 3.6].

(a) By Lemma 2.12(b), we can write

$$\underline{p} \cdot A_E^{-1} \cdot \underline{p} = A\binom{i}{i}^{-1} \sum_{j,k \neq i} p_j A\binom{i\,j}{i\,k} p_k \quad , \tag{3.13}$$

and by Lemmas A.11 and A.13,

$$\alpha_1 \cdots \alpha_L A\binom{i\,j}{i\,k} = \sum_{T_2} \prod_{\ell \notin T_2} \alpha_\ell \quad , \tag{3.14}$$

the sum taken over all 2-trees $T_2$ for which i lies in a different component from j and k. We argue just as in the proof of Lemma 3.6 that when we make the variable change (3.10), each term in the sum in (3.14) contains a factor $\Pi_{\mathfrak{S}}\, t_G^{N(G)}$. We may cancel this factor from the numerator and denominator of (3.13), along with Lemma 3.6, this completes the proof.

(b) From (3.3a), Lemma 2.12(b) (taking $j = i_\ell$), and Lemmas A.11 and A.13,

$$\begin{aligned} Y^{(\ell)} &= \frac{1}{\alpha_\ell} \sum_{r \neq i_\ell} \frac{A(i_\ell \mid f_\ell\, r)}{A\binom{i}{i}}\, p_r \\ &= \frac{1}{\alpha_\ell} \sum_{r \neq i_\ell} \left[ \sum_{T_2} \prod_{\ell' \notin T_2} \alpha_{\ell'} \right] \frac{p_r}{[\alpha_1 \cdots \alpha_L\, A\binom{i}{i}]} \quad , \end{aligned} \tag{3.15}$$

where $\Sigma_{T_2}$ runs over all 2-trees with $i_\ell$ in a different component from $f_\ell$ and r. But then $T_2 \cup \{\ell\}$ is a tree, so

$$\frac{1}{\alpha_\ell} \left[ \sum_{T_2} \prod_{\ell' \notin T_2} \alpha_{\ell'} \right] = \sum_T{}' \prod_{\ell' \notin T} \alpha_{\ell'} \tag{3.16}$$

where $\Sigma'_T$ runs over certain trees of $G_0$. Again, as in Lemma 3.6, (3.16) contains the factor $\Pi_{\mathfrak{S}}\, t_G^{N(G)}$, which may be cancelled in (3.15).

(c) Take $\ell \neq \ell'$. Arguing as in (b), we have

$$\overline{Y^{(\ell)}Y^{(\ell')}} = -\frac{i}{2}\,\frac{A(i_\ell|f_\ell f_{\ell'}) - A(i_\ell\,|\,f_\ell i_{\ell'})}{\alpha_\ell\,\alpha_{\ell'}\,A(\tfrac{i}{i})}$$

$$= -\frac{i}{2}\,\frac{A(i_\ell i_{\ell'}\,|\,f_\ell f_{\ell'}) - A(i_\ell f_{\ell'}|\,f_\ell i_{\ell'})}{\alpha_\ell\,\alpha_{\ell'}\,A(\tfrac{i}{i})}\ ,$$

where we have used (A.6) in the second step. Thus $\overline{Y^{(\ell)}Y^{(\ell')}}$ is the difference of two terms of the form

$$-\frac{i}{2\alpha_\ell\alpha_{\ell'}}\,\frac{\Sigma_{T_2}\,\Pi_{\ell''\notin T_2}\,\alpha_{\ell''}}{(\alpha_1\,\cdots\,\alpha_L)\,A(\tfrac{i}{i})}\ , \tag{3.17}$$

where each $T_2$ in the sum in (3.17) is such that both $T_2 \cup \{\ell\}$ and $T_2 \cup \{\ell'\}$ are trees. Thus $\{\ell''|\ell''\notin T_2\}$ contains $\ell$ and $\ell'$; for each $G \in \mathcal{E}$ it contains $N(G)$ lines from $G$, and if $\ell \in G$ or $\ell' \in G$, it contains $N(G)+1$ lines from $G$. That is, the numerator in (3.17) has the form

$$\beta_\ell\beta_{\ell'}\,\prod_{\mathcal{E}} t_G^{N(G)}\ \prod_{G\ni\ell\ \text{or}\ G\ni\ell'} t_G\ \text{[polynomial]}\ ,$$

and using $\alpha_\ell = \beta_\ell\,\Pi_{G\ni\ell}\,t_G$ and Lemma 3.6, (3.17) becomes

$$-\frac{i}{2}\prod_{G\ni\ell\ \textit{and}\ G\ni\ell'} t_G^{-1}\ \frac{\text{[polynomial]}}{B(\underline{t},\underline{\beta})}\ .$$

This proves (c).

(d) Again arguing as in (b), we have

$$\overline{Y^{(\ell)}Y^{(\ell)}} = -\frac{i}{2\alpha_\ell^2\,A(\tfrac{i}{i})}\,[A(i_\ell\,|\,f_\ell) - \alpha_\ell\,A(\tfrac{i}{i})]\ .$$

But

$$\alpha_1\,\cdots\,\alpha_L\,A(i_\ell\,|\,f_\ell) = \sum_{T_2}\ \prod_{\ell'\notin T_2}\alpha_{\ell'}\ ,$$

the sum taken over all 2-trees $T_2$ such that $T_2 \cup \{\ell\}$ is a tree. Thus

$$\overline{Y^{(\ell)}Y^{(\ell)}} = \frac{-i}{2\alpha_\ell[\alpha_1\cdots\alpha_L]\,A(\tfrac{i}{i})}\left[\sum_{T\ni\ell}\ \prod_{\ell'\notin T}\alpha_\ell - (\alpha_1\cdots\alpha_L)\,A(\tfrac{i}{i})\right]$$

$$= \frac{-i}{2[\alpha_1\cdots\alpha_L]\,A(\tfrac{i}{i})}\left[\sum_{T^*}\ \prod_{\ell'\notin T^*}\alpha_\ell\right]\ ,$$

where $\Sigma_{T^*}$ runs over all sets of the form $T \cup \{\ell\}$, with $T$ a tree and $\ell \notin T$. Then for each $G \in \mathcal{E}$, $\{\ell'|\ell'\notin T^*\}$ contains at least $N(G)$ [respectively $N(G)-1$] lines from $G$

if G does not [respectively does] contain $\ell$. Thus the $\Sigma_{T^*}$ in (3.18) has the form

$$\prod_{G \ni \ell} t_G^{-1} \prod_{\mathcal{E}} t_G^{N(G)} \text{ [polynomial] },$$

and with Lemma 3.6 this proves (d).

Now we proceed with the proof of Theorem 3.4. When we expand the $\tau$-product in (3.9) we obtain terms of the form

$$\prod_{\ell \in \mathfrak{L}} [Y^{(\ell)}]^{a_\ell} \prod_{\ell \in \mathfrak{L}} [\overline{Y^{(\ell)} Y^{(\ell)}}]^{b_\ell} \prod_{\substack{\ell,\ell' \in \mathfrak{L} \\ \ell \neq \ell'}} [\overline{Y^{(\ell)} Y^{(\ell')}}]^{c_{\ell,\ell'}}, \tag{3.19}$$

where $a_\ell$, $b_\ell$, and $c_{\ell,\ell'}$ are integers satisfying

$$a_\ell + 2b_\ell + \sum_{\ell' \neq \ell} (c_{\ell,\ell'} + c_{\ell',\ell}) \leq r_\ell . \tag{3.20}$$

Now make the change of variables (3.10) in the integral (3.9); the Jacobian is $\prod_{\mathcal{E}} t_G^{L(G)-1}$. Thus using (3.19) and Lemmas 3.6 and 3.7, $\mathcal{T}_\epsilon(\underline{\lambda}, \mathcal{E}, \sigma)$ becomes a sum of terms of the form

$$\begin{aligned} & i\, b(\underline{\lambda})\, \delta\Big(\sum_1^n p_i\Big) \int_{\mathfrak{D}'(\mathcal{E},\sigma)} [\prod \gamma_\ell^{\lambda_\ell - 1}\, d\gamma_\ell][\prod \beta_\ell^{\lambda_\ell - 1}\, d\beta_\ell] \times \\ & \qquad \times \prod_{\mathcal{E}} t_G^{\nu(G)-1} F(\underline{\gamma}, \underline{\beta}, \underline{t}) , \end{aligned} \tag{3.21}$$

where

$$\nu(G) = \sum_{\ell \in \mathfrak{L}(G)} \lambda_\ell - 2N(G) - \sum_{\ell \in \mathfrak{L}(G)} b_\ell - \sum_{\substack{\ell,\ell' \in \mathfrak{L}(G) \\ \ell \neq \ell'}} c_{\ell,\ell'} , \tag{3.22}$$

and $F(\underline{\gamma}, \underline{\beta}, \underline{t})$ is $C^\infty$ in $\mathfrak{D}'(\mathcal{E},\sigma)$ and decreases exponentially as $\gamma_\ell$ and $t_{H_i}$ approach infinity.

Now (3.21) is a convergent integral, analytic in $\underline{\lambda}$, as long as $\operatorname{Re}\lambda_\ell > 0$ and $\operatorname{Re}\nu(G) > 0$ for any $\ell \in T_0$ and $G \in \mathcal{E}$. We simply integrate by parts with respect to $\gamma_\ell$, $\beta_\ell$, and $t_G$, integrating the factors $\gamma_\ell^{\lambda_\ell - 1}$, $\beta_\ell^{\lambda_\ell - 1}$, and $t_G^{\nu(G)-1}$ (or the higher powers arising from them) and repeatedly differentiating $F(\underline{\gamma}, \underline{\beta}, \underline{t})$. If we integrate $\gamma_\ell$ and $\beta_\ell$ by parts $m_\ell$ times, and $t_G$ $m_G$ times, we produce new integrals analytic in $\underline{\lambda}$ in the region

$$\begin{aligned} \operatorname{Re}\lambda_\ell &> m_\ell \quad (\ell \in T_0) , \\ \operatorname{Re}\nu(G) &> m_G \quad (G \in \mathcal{E}) . \end{aligned} \tag{3.23}$$

Clearly the region (3.23) expands to fill all of $\mathbb{C}^L$ as $m_\ell$ and $m_G$ approach infinity. However, the integrals produced are multiplied by the factor

$$\prod_{\ell \in T_0} \left( \frac{1}{\lambda_\ell} \frac{1}{\lambda_\ell + 1} \cdots \frac{1}{\lambda_\ell + m_\ell} \right) \prod_{G \in \mathcal{E}} \left( \frac{1}{\nu(G)} \cdots \frac{1}{\nu(G) + m_G - 1} \right) \tag{3.24}$$

[Actually, this is not quite true. Integrating (3.21) by parts actually produces a sum of integrals, some of which correspond to evaluations at $\beta_\ell = 1$ or $t_G = 1$ during a $\beta_\ell$ or $t_G$ integration. These integrals will be multiplied by an expression similar to (3.24) but with some factors missing. This does not affect the following discussion.] Thus we see that the integral in (3.21), if multiplied by

$$\prod_{T_0} [\Gamma(\lambda_\ell)]^{-1} \prod_{\mathfrak{S}} \{\Gamma[\nu(G)]\}^{-1} ,$$

has an analytic extension to all of $C^L$. By (3.4), however, the factors $\Pi_{T_0}[\Gamma(\lambda_\ell)]^{-1}$ already occur in $b(\underline{\lambda})$. We conclude: the expression (3.21), multiplied by

$$\prod_{G \in \mathfrak{S}} \{\Gamma[\nu(G)]\}^{-1} , \tag{3.25}$$

has an analytic extension to $C^L$.

This almost completes the proof of Theorem 3.4; it remains only to relate (3.25) to the factors occurring in (3.7). Specifically, we show that for $G \in \mathfrak{S}$,

$$\{\Gamma(\Sigma_{\ell \in G} \lambda_\ell - L(G) - [\tfrac{\mu(G)}{2}])\}^{-1} \Gamma[\nu(G)] \tag{3.26}$$

is entire in $\underline{\lambda}$. From Definition 1.4 we have

$$2N(G) = \frac{1}{2}\mu(G) + L(G) - \sum_{\ell \in \mathfrak{L}(G)} \frac{r_\ell}{2} , \tag{3.27}$$

while using (3.20),

$$\begin{aligned} \sum_{\ell \in \mathfrak{L}(G)} b_\ell + \sum_{\substack{\ell,\ell' \in \mathfrak{L}(G) \\ \ell \neq \ell'}} c_{\ell,\ell'} &= \sum_{\ell \in \mathfrak{L}(G)} b_\ell + \sum_{\ell \in \mathfrak{L}(G)} \left[ \frac{1}{2} \sum_{\substack{\ell' \in \mathfrak{L}(G) \\ \ell' \neq \ell}} c_{\ell,\ell'} + c_{\ell',\ell} \right] \\ &\leq \sum_{\ell \in \mathfrak{L}(G)} \frac{r_\ell}{2} . \end{aligned} \tag{3.28}$$

Adding (3.27) and (3.28), and using the fact that the left hand side of the resulting inequality is an integer, we have

$$2N(G) + \sum_{\ell \in \mathfrak{L}(G)} b_\ell + \sum_{\substack{\ell,\ell' \in \mathfrak{L}(G) \\ \ell \neq \ell'}} c_{\ell,\ell'} \leq [\tfrac{1}{2}\mu(G)] + L(G) ,$$

and, combined with (3.22), this shows the analyticity of (3.26).

We now discuss the $\epsilon \to 0$ limit of $\mathcal{T}_\epsilon(\underline{\lambda})$.

*Theorem 3.8:* Let $K \subset C^L$ be a compact set which does not contain any singularities of $\mathcal{T}_\epsilon(\underline{\lambda})$. Then the limit

$$\mathcal{T}(\underline{\lambda}) \doteq \lim_{\epsilon \to 0} \mathcal{T}_\epsilon(\underline{\lambda})$$

exists for $\underline{\lambda} \in K$, uniformly for some norm in $\mathcal{S}(R^{4n})$. Thus $\mathcal{T}(\underline{\lambda})$ is a distribution analytic in $\underline{\lambda}$ with the same singularities as $\mathcal{T}_\epsilon(\underline{\lambda})$.

*Proof:* We get a representation of $\mathcal{T}_\epsilon(\underline{\lambda})$ valid for $\underline{\lambda} \in K$ by writing $\mathcal{T}_\epsilon(\underline{\lambda})$ as a sum of terms of the form (3.21), and then doing enough integrations by parts so that the region (3.23) contains K. Thus $\mathcal{T}_\epsilon(\underline{\lambda})$ is a sum of terms of the form

$$H(\underline{\lambda})\,\delta\Big(\sum_{i=1}^{n} p_i\Big) \int_0^1 \cdots \int_0^1 \prod_{\ell \in \mathcal{L}_1} \beta_\ell^{\rho_\ell - 1} d\beta_\ell \int_0^\infty \cdots \int_0^\infty \prod_{\ell \in \mathcal{L}-\mathcal{L}'} \gamma_\ell^{\rho_\ell - 1} d\gamma_\ell \int_0^1 \cdots \int_0^1 \prod_{G \in \mathcal{G}_1} t_G^{\rho_G - 1} dt_G$$

$$\int_0^\infty \cdots \int_0^\infty \prod_{i=1}^{k} t_{H_i}^{\rho_{H_i} - 1} dt_{H_i}\, F'(\underline{\gamma}, \underline{\beta}, \underline{t}) \exp i[\underline{p} \cdot A_E^{-1} \cdot \underline{p} - \sum_{\mathcal{L}} a_\ell (m_\ell^2 - i\epsilon)] \;. \tag{3.29}$$

Here $H(\underline{\lambda})$ may contain singularities like (3.24), $\mathcal{L}_1 \subset (\mathcal{L}' - \sigma(\mathcal{G}))$, and $\mathcal{G}_1 \subset [\mathcal{G} - \{H_1, \ldots, H_k\}]$ (note that some $\beta_\ell$ and $t_G$ may have been set equal to one during a partial integration). Now $F'(\underline{\gamma}, \underline{\beta}, \underline{t})$ may be taken homogeneous of some degree, say m, in the set of variables $\{\gamma_\ell, t_{H_i}\}$; $A_E^{-1}$ is homogeneous of degree one in these variables, as are $a_1, \ldots, a_L$. Thus if we change variables in (3.29) to $\{u, \gamma_\ell', \beta_\ell, t_G, t_{H_i}'\}$ given by

$$u = \sum_{\mathcal{L}-\mathcal{L}'} \gamma_\ell + \sum_{i=1}^{k} t_{H_i} \;,$$

$$\gamma_\ell = u\gamma_\ell' \;,$$

$$t_{H_i} = u t_{H_i}' \;,$$

and write $u a_\ell' = a_\ell$, we may do the u-integral in (3.29) explicitly, using Lemma 2.8 (a), to get

$$H(\underline{\lambda})\,\delta\Big(\sum_{i=1}^{n} p_i\Big) \int_0^1 \cdots \int_0^1 \prod_{\mathcal{L}_1} \beta_\ell^{\rho_\ell - 1} d\beta_\ell \prod_{\mathcal{L}-\mathcal{L}'} \gamma_\ell'^{\,\rho_\ell - 1} d\gamma_\ell' \prod_{\mathcal{G}_1} t_G^{\rho_G - 1} dt_G\, \delta(1 - \Sigma\gamma_\ell' - \Sigma t_{H_i}')$$

$$\prod_{i=1}^{k} t_{H_i}'^{\,\rho_{H_i} - 1} dt_{H_i}'\, F'(\underline{\gamma}', \underline{\beta}, \underline{t}')\,\Gamma(\nu)\, e^{\nu\pi i/2} [\underline{p} \cdot A_E^{-1\prime} \cdot \underline{p} - \Sigma\, a_\ell'(m_\ell^2 - i\epsilon)]^{-\nu} \tag{3.30}$$

where

$$\nu = \Big[\sum_{i=1}^{k} \rho_{H_i} + \sum_{\mathcal{L}-\mathcal{L}'} \rho_\ell + m\Big] \;,$$

and we have written $A_E^{-1\prime}$ to indicate that $\gamma_\ell$ and $t_{H_i}$ are to be replaced by $\gamma_\ell'$ and $t_{H_i}'$.

But by Theorem B.8 the distribution $[\underline{p}\cdot A_E^{-1'}\cdot\underline{p} - \Sigma\, \alpha_\ell'(m_\ell^2 - i\epsilon)]^{-\nu}$ has an $\epsilon \to 0$ limit uniformly [in some norm on $\mathcal{S}(R^{4n})$] when the parameters vary over the compact region of integration in (3.30); this proves the theorem.

*Remark 3.9:* Recall from Chapter I that the Feynman amplitude in the perturbation series was integrated over the x variables corresponding to internal vertices. This is equivalent to setting the corresponding momentum variables equal to zero, that is, restricting the distribution (3.30) to a subspace of $R^{4n}$ of the form $\{p_i^\mu = 0 \mid i = i_1, i_2, \ldots, i_k\}$. This is clearly permissible as long as at least one $p_i$ is not set to zero (in which case the $\delta(\Sigma\, p_i)$ would give a divergence). Since our amplitudes in Chapter I all had at least one external leg, the x-integrations there give well defined distributions in the external variables.

Section 3. REMOVING THE $\underline{\lambda}$ -SINGULARITY.

We have now defined the GFA $\mathcal{T}(\underline{\lambda})$ for all $\underline{\lambda} \in C^L$. It corresponds formally to the true Feynman amplitude of $G_0$ for $\lambda_\ell = 1$; however, we have seen (Remark 3.5) that the divergence of the Feynman integrals, with the consequent need for renormalization, corresponds precisely to the existence of a singularity of $\mathcal{T}(\underline{\lambda})$ at $\lambda_\ell = 1$. In this section we use the analytic properties (in $\underline{\lambda}$) of $\mathcal{T}(\underline{\lambda})$ to "subtract" the singularity at $\lambda_\ell = 1$, producing a "finite part." In the next section we show that this finite part is a renormalized amplitude for $G_0$ according to Definition 1.19.

Now if $\mathcal{T}$ depended on only one complex variable $z$ and had an isolated singularity at $z = z_0$, it would be natural to define the finite part at $z_0$ to be the constant term of the Laurent series of $\mathcal{T}$ about $z_0$. The question then arises: why not set $\lambda_1 = \cdots = \lambda_L = z$ in $\mathcal{T}(\lambda)$ and apply the above prescription at $z = 1$? We give an example to show why this is unsatisfactory.

*Example 3.10:* Let $G_0$ and $G_1$ be the graphs of Figure 3.1, with $\mathcal{T}_0$ and $\mathcal{T}_1$ their GFA's.

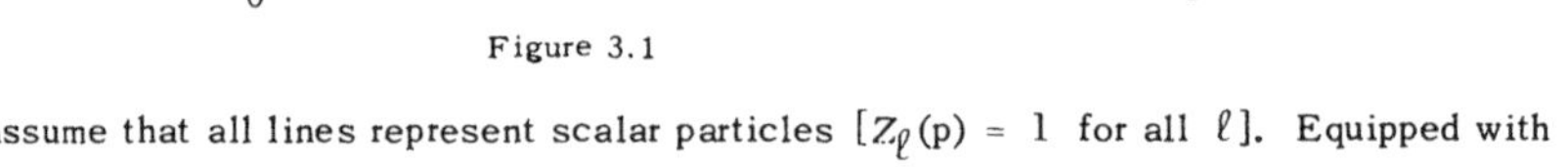

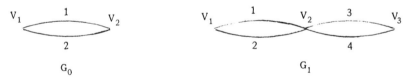

Figure 3.1

We assume that all lines represent scalar particles [$Z_\ell(p) = 1$ for all $\ell$]. Equipped with two external lines at each end, for example, $G_0$ and $G_1$ could be scattering diagrams in a $\phi^4$ theory. Now if we write $\xi_1 = x_2 - x_1$, $\xi_2 = x_3 - x_2$, then $\mathcal{T}_0$ has the form

$$\mathcal{T}_0(\lambda_1, \lambda_2)(\underline{x}) = \mathcal{T}_0'(\lambda_1, \lambda_2)(\xi_1) \,, \tag{3.31}$$

and we also have

$$\mathcal{T}_1(\lambda_1, \lambda_2, \lambda_3, \lambda_4)(\underline{x}) = \mathcal{T}_0'(\lambda_1, \lambda_2)(\xi_1)\,\mathcal{T}_0'(\lambda_3, \lambda_4)(\xi_2) \; . \tag{3.32}$$

The product in (3.32) is a tensor product and hence well defined.

Now set $\lambda_1 = \lambda_2 = \lambda_3 = \lambda_4 = z$. From Theorem 3.4, $\mathcal{T}_0$ has a simple pole at $z = 1$, so the Laurent series of (3.31) has the form

$$\mathcal{T}_0'(z, z)(\xi_1) = \frac{a}{z-1} + b(\xi_1) + (z-1)\,c(\xi_1) + \cdots \; . \tag{3.33}$$

Thus our candidate for a finite part of $\mathcal{T}_0$ is $b(\xi_1)$. Now it is easily shown that $b(\xi_1)$ is indeed a correctly renormalized amplitude for the graph $G_0$; that is, our prescription is satisfactory in this case (this is an easy calculation from Definition 1.19 and (3.2); one calculates similarly that $a$ in (3.33) is independent of $\xi_1$. See also [30]).

The difficulty arises when we pass to $G_1$. Definition 1.19 implies that if $b(\xi_1)$ is the renormalized amplitude for $G_0$, then the renormalized amplitude for $G_1$ must have the form

$$b(\xi_1)\,b(\xi_2) + d \; , \tag{3.34}$$

where $d$ is a constant arising from the $\hat{\mathfrak{X}}$ term in (1.81). From (3.32) and (3.33), however, the Laurent series of $\mathcal{T}_1$ is

$$\mathcal{T}(z, z, z, z)(\underline{x}) = \frac{a^2}{(z-1)^2} + \frac{a[b(\xi_1) + b(\xi_2)]}{(z-1)} + \{b(\xi_1)\,b(\xi_2) + a[c(\xi_1) + c(\xi_2)]\} + \cdots \; , \tag{3.35}$$

and the constant term of (3.35),

$$b(\xi_1)\;b(\xi_2) + a[c(\xi_1) + c(\xi_2)] \; , \tag{3.36}$$

does not agree with (3.34) because (as may be calculated) $a$ is not zero and $c$ is not independent of $\xi_1$. Thus our prescription fails for the graph $G_1$.

We now define a class of operations which specify a finite part for $\mathcal{T}(\underline{\lambda})$ in a suitable way. Note that the difficulty in Example 3.10 arose because a pole in $\underline{\lambda}$ associated with one subgraph of the diagram, produced unwanted higher $\lambda$-derivatives in the amplitude for the rest of the graph (specifically, the terms $ac(\xi_1)$, $ac(\xi_2)$ in (3.36)). This happened because we set all the $\lambda_\ell$'s equal, thus losing the knowledge of where the pole arose. It is the desire to avoid this behavior which motivates the important condition (W6) of Definition 3.11.

*Definition 3.11:* For each $L \geq 1$, set

$$j_L(\lambda_1, \ldots, \lambda_L) = \prod_K \left[ \sum_{\ell \in K} (\lambda_\ell - 1) \right] \; ,$$

where $\Pi_K$ runs over all nonempty $K \subset \{1, \ldots, L\}$. For $\epsilon > 0$, define $U_{L,\epsilon} = \{(\lambda_1, \ldots, \lambda_L) \mid |\lambda_\ell - 1| < \epsilon,\ \ell = 1, \ldots, L\}$, and $\mathfrak{A}_{L,\epsilon} = \{f(\lambda_1, \ldots, \lambda_L) \mid f(\underline{\lambda})\, j_L(\underline{\lambda}) \text{ is analytic in } U_{L,\epsilon}\}$.

When $\epsilon' < \epsilon$ we have the inclusion $\mathcal{A}_{L,\epsilon} \subset \mathcal{A}_{L,\epsilon'}$, thus we may define $\mathcal{A}_L = \cup_{\epsilon>0} \mathcal{A}_{L,\epsilon}$. Finally, $\mathcal{A}_{L,\epsilon}$ is topologized by the family of norms

$$\|f\|_m = \sup_{\underline{\lambda} \in U_{L,(1-\frac{1}{m})\epsilon}} |f(\underline{\lambda})\, j_L(\underline{\lambda})| \quad ,$$

that is, $\mathcal{A}_{L,\epsilon}$ has the topology of uniform convergence on compact sets of the products $f(\underline{\lambda})|j_L(\underline{\lambda})$.

Then a family of maps $\mathcal{W} = \{\mathcal{W}_L \mid L = 1, 2, \ldots\}$, with $\mathcal{W}_L : \mathcal{A}_L \to C$, is a *generalized evaluator* if the following conditions are satisfied for each $L$:

(W 1). $\mathcal{W}_L$ is linear;

(W 2). If $f \in \mathcal{A}_L$ is analytic at $(1, \ldots, 1)$, then $\mathcal{W}_L f = f(1, \ldots, 1)$;

(W 3). $\mathcal{W}_L$ is continuous on $\mathcal{A}_{L,\epsilon}$ for any $\epsilon > 0$;

(W 4). If $s$ is a permutation on $\{1, \ldots, L\}$, $f \in \mathcal{A}_L$, and $f_s \in \mathcal{A}_L$ is defined by $f_s(\lambda_1, \ldots, \lambda_L) = f(\lambda_{s(1)}, \ldots, \lambda_{s(L)})$, then $\mathcal{W}_L f_s = \mathcal{W}_L f$;

(W 5). If $f \in \mathcal{A}_L$ is independent of $\lambda_{L'+1}, \ldots, \lambda_L$ for some $L' < L$, then $\mathcal{W}_{L'} f = \mathcal{W}_L f$;

(W 6). If $f_1, f_2 \in \mathcal{A}_L$, and $f_1$ is independent of $\lambda_1, \ldots, \lambda_{L'}$, $f_2$ independent of $\lambda_{L'+1}, \ldots, \lambda_L$, then $\mathcal{W}_L(f_1 f_2) = (\mathcal{W}_L f_1)(\mathcal{W}_L f_2)$.

*Remark 3.12* (a). If $f \in \mathcal{A}_L$, we use (W 4) and (W 5) to write without ambiguity $\mathcal{W} f = \mathcal{W}_L f = \mathcal{W}_{L'} f$ for any $L' \geq L$. (b). We need to be able to apply a generalized evaluator to meromorphic distributions. Consider such a distribution $S(\underline{\lambda}) = S'(\underline{\lambda})\, j_L(\underline{\lambda})$, where $S'(\underline{\lambda})$ is analytic in $U_{L,\epsilon}$ for some $\epsilon > 0$. The formula

$$[\mathcal{W}_L S](\psi) = \mathcal{W}_L[S(\psi)]$$

defines a linear functional $\mathcal{W}_L S$ on $\mathcal{S}(R^{4n})$; to show $\mathcal{W}_L S \in \mathcal{S}'(R^{4n})$ we must verify continuity. Now $S' : U_\epsilon \to \mathcal{S}'(R^{4n})$ is continuous, so that if $K \subset U_\epsilon$ is compact, $S'(K)$ is compact and hence uniformly bounded [13, Chap. 1, Sec. 5; see also Definition B.1]. Thus for any sequence $\{\psi_i \mid \psi_i \in \mathcal{S}(R^{4n})\}$ with $\psi_i \to \psi_0$, the sequence $\{S'(\underline{\lambda})(\psi_i)\}$ converges uniformly on $K$ to $S'(\underline{\lambda})(\psi_0)$; then (W 3) implies that $\mathcal{W}_L S$ is continuous. (c). It is instructive to apply $\mathcal{W}$ to the amplitude (3.32); one sees easily that (W 1)–(W 6) guarantee that $\mathcal{W}\mathcal{T}_1$ has the form (3.34).

*Example 3.13:* We give an example of a generalized evaluator. Suppose $f \in \mathcal{A}_{L,\epsilon}$, and choose $0 < R_1 < \cdots < R_L < \epsilon$ so that $R_i > \Sigma_{j=1}^{i-1} R_j$. Let $C_i$ be the contour $|z-1| = R_i$ oriented counterclockwise. Then

$$\mathcal{W}_L f = \frac{1}{L!} \sum_{s \in S_L} \frac{1}{(2\pi i)^L} \int_{C_{s(1)}} d\lambda_1 \cdots \int_{C_{s(L)}} d\lambda_L \frac{f(\underline{\lambda})}{(\lambda_1 - 1) \cdots (\lambda_L - 1)} \quad .$$

One easily checks that $\mathcal{W}$ is independent of the choice of $R_1, \ldots, R_L$ and satisfies (W 1)–(W 6).

*Definition 3.14:* Let $\mathcal{W}$ be a generalized evaluator, $\mathcal{T}(\underline{\lambda})$ the GFA for some graph $G_0$. The *analytically renormalized Feynman amplitude* for $G_0$ is given by $\mathcal{W}\mathcal{T}$.

*Remark 3.15:* (a). This definition is justified because Theorems 3.4 and 3.8 show that $j_L(\underline{\lambda})\mathcal{T}(\underline{\lambda})$ is analytic at $\lambda_\ell = 1$ (here $L = \#[\mathcal{L}(G_0)]$).
(b). By Theorem 3.4 we may also apply $\mathcal{R}$ to $\mathcal{T}_\epsilon(\underline{\lambda})$ for $\epsilon > 0$; then from Theorem 3.9 and (W 3),

$$\mathcal{R}\mathcal{T} = \lim_{\epsilon \to 0} \mathcal{R}\mathcal{T}_\epsilon .$$

Section 4. VALIDITY OF ANALYTIC RENORMALIZATION.

In this section we prove the fact that Definition 3.14 is a special case of Definition 1.19, i.e., that analytic renormalization is indeed renormalization and not some other form of regularization. Since we will have to refer to the GFA of various subgraphs of our basic graph $G_0$, we adopt notation uniform with that of Chapter I; specifically, if $V_1', \ldots, V_m'$ are vertices of $G_0$, then $\mathcal{T}(\underline{\lambda}; V_1', \ldots, V_m')$ is the GFA of the graph $G(V_1', \ldots, V_m')$ (Definition A.2). [Actually, $\mathcal{T}(\underline{\lambda}; V_1', \ldots, V_m')$ depends only on these $\lambda_\ell$ for which both $V_{i_\ell}$ and $V_{f_\ell}$ lie in $\{V_1', \ldots, V_m'\}$, but we write it as a function of $\underline{\lambda} = (\lambda_1, \ldots, \lambda_L)$.]

We will need the relation between the GFA $\mathcal{T}(\underline{\lambda})$ and the quantities defined in Chapter I.

*Lemma 3.16:* Let $\mathcal{T}_{\epsilon,r}(\underline{\lambda})$ (with $r > 0$) be defined by (3.2), modified so that the lower limit on all $\alpha_\ell$ integrals is $r$. Then

$$\mathcal{T}_{\epsilon,r}(\underline{\lambda})(\underline{x}) = \prod_{\mathcal{L}} \Delta^{(\ell)}_{\epsilon,r}(\lambda_\ell)(x_{f_\ell} - x_{i_\ell}) . \tag{3.37}$$

[$\Delta^{(\ell)}_{\epsilon,r}(\lambda_\ell)$ is defined in Definition 1.20.]

*Proof:* We omit details of the proof, simply noting that the right hand side of (3.37) may be calculated explicitly from Definition 1.20, using the same techniques as were used in Chapter II. This will give (3.37).

*Remark 3.17:* In Chapter I, Section 2.B, we gave formula (1.43) for $\mathcal{T}_{\epsilon,r}(\underline{x}) = \prod_{\mathcal{L}} \Delta^{(\ell)}_{\epsilon,r}(x_{f_\ell} - x_{i_\ell})$. This may now be justified as follows: From Lemma 3.16 we have

$$\mathcal{T}_{\epsilon,r} = \mathcal{T}_{\epsilon,r}(\underline{\lambda})\Big|_{\lambda_\ell = 1} .$$

Now (1.43) follows from (3.2), taking
(a) $C(\underline{\alpha}) = \alpha_1 \alpha_2 \cdots \alpha_L\, A(\begin{smallmatrix} i \\ i \end{smallmatrix})$;
(b) $B(\underline{\alpha}, \underline{p}) = i\,b(1, 1 \cdots 1)\, r[\prod_{\mathcal{L}} Z_\ell(Y^{(\ell)})]$;
(c) $D(\underline{\alpha}, \underline{p}) = \alpha_1 \alpha_2 \cdots \alpha_L \sum_{k,j \neq i} p_k\, A(\begin{smallmatrix} i & j \\ i & k \end{smallmatrix})\, p_j$.

All the distributions that concern us here belong to a certain class $\mathcal{B}(L, m)$, which we now define and discuss.

*Definition 3.18:* Let $\mu = \max \left[ \frac{\mu(G)}{2} \right]$, the maximum taken over all subgraphs G of $G_0$, and let

$$J_L(\lambda_1, \ldots, \lambda_L) = \prod_K \Gamma\left[ \sum_{\ell \in K} \lambda_\ell - L - \mu \right] ,$$

where $\Pi_K$ runs over all nonempty $K \subset \{1, \ldots, L\}$ (compare Definition 3.11). Let $\mathcal{B}(L, m)$ be the set of all mappings $\phi: C^L \to \mathcal{S}'(R^{4m})$ with the form

$$\phi(\underline{\lambda})(p_1, \ldots, p_m) = \delta\left( \sum_{i=1}^{m} p_i \right) J_L(\underline{\lambda}) f(\underline{\lambda}, p_1, \ldots, p_m), \tag{3.38}$$

where (a) $f \in C^\infty(R^{2L+4m})$ ,

(b) f is analytic in $\underline{\lambda}$ for fixed $p_1, \ldots, p_m$,

(c) if D is a monomial in the p-derivatives, and $X \subset C^L$ a compact set, there are positive constants $C_1$, $C_2$ such that

$$|(Df)(\underline{\lambda}, p_1, \ldots, p_m)| \le C_1 (1 + \|p\|^2)^{C_2} ,$$

uniformly for $\underline{\lambda} \in X$.

Finally, for any integer $\nu$, define $\mathcal{M}_\nu$: $\mathcal{B}(L, m) \to \mathcal{B}(L, m)$ by

$$[\mathcal{M}_\nu \phi](\underline{\lambda})(p_1, \ldots, p_m) = \delta\left( \sum_1^m p_i \right) J_L(\underline{\lambda}) F_\nu(\underline{\lambda}, p_1, \ldots, p_m) ,$$

where $\phi$ is given by (3.38) and $F_\nu$ is the Taylor series of F in p about the origin up to order $\nu$ (with $\mathcal{M}_\nu = 0$ if $\nu < 0$). (Compare Definition 1.3).

*Lemma 3.19:* Let $\mathcal{W}$ be a generalized evaluator. Then $\mathcal{W}$ maps $\mathcal{B}(L, m)$ into $\mathcal{B}(L, m)$ and commutes with $\mathcal{M}_\nu$ (for any $\nu$) on $\mathcal{B}(L, m)$.

*Proof:* $\mathcal{W}$ is defined on an element $\phi \in \mathcal{B}(L, m)$ by $(\mathcal{W}\phi)(\psi) = \mathcal{W}[\phi(\psi)]$, for any $\psi \in \mathcal{S}(R^{4m})$ [Remark 3.12(b)]. We claim that, if $\phi$ is given by (3.38),

$$[\mathcal{W}\phi](\underline{p}) = \delta\left( \sum_1^m p_i \right) \mathcal{W}[J(\underline{\lambda}) f(\underline{\lambda}, \underline{p})] . \tag{3.39}$$

Note first that the difference quotient defining a p-derivative of f converges uniformly for $\underline{\lambda}$ in a compact set, so that (W 3) implies that $\mathcal{W}[J_L(\underline{\lambda}) f(\underline{\lambda}, \underline{p})] \in C^\infty(R^{4m})$. Moreover, Definition 3.18 (c) implies that for $\underline{\lambda} \in X$, the quantity

$$f(\underline{\lambda}, \underline{p})(1 + \|p\|^2)^{-(C_2+1)}$$

approaches zero uniformly as $\|p\| \to \infty$; thus (W 3) implies $\mathcal{W}[J_L(\underline{\lambda}) f(\underline{\lambda}, \underline{p})] \in \mathcal{O}_M(R^{4m})$ [31], that is, (3.39) is indeed in $\mathcal{B}(L, m)$ (as a constant function of $\underline{\lambda}$). Now

$$\phi(\underline{\lambda})(\psi) = \int_E \psi(\underline{p}) J_L(\underline{\lambda}) f(\underline{\lambda}, \underline{p}) d\underline{p} ,$$

and this integral may be approximated, uniformly in compact subsets of $\mathbb{C}^L$, by Riemann sums. Then (W 1) and (W 3) imply (3.39). The fact that $\mathcal{M}_\nu$ and $\mathcal{W}$ commute follows from the uniformity in $\underline{\lambda}$ of the limit defining a p-derivative.

Now Theorem 3.4 implies that $\mathcal{T}_\epsilon(\underline{\lambda};\ V_1' \cdots V_m') \in \mathcal{B}(L, m)$ for any $\{V_1' \cdots V_m'\} \subset \{V_1 \cdots V_n\}$. Thus we may define

$$(3.40) \qquad \hat{\mathcal{X}}_\epsilon(V_1' \cdots V_m') = \begin{cases} 1, & \text{if } m = 1; \\ 0, & \text{for IPR } G(V_1' \cdots V_m'); \\ \mathcal{W}\mathcal{M}_{\mu(V_1' \cdots V_m')}\mathcal{T}_\epsilon(\underline{\lambda};\ V_1' \cdots V_m') & \text{otherwise}. \end{cases}$$

*Lemma 3.20:* $\hat{\mathcal{X}}_\epsilon$ as given by (3.40) is a finite renormalization (Definition 1.18), moreover, $\hat{\mathcal{X}}_\epsilon(V_1' \cdots V_m')$ depends only on the structure of $G(V_1', \ldots, V_m')$. (Recall from Chapter I, Section 2.E that the latter statement characterized those finite renormalizations which could be incorporated into a field theory.)

*Proof:* $\hat{\mathcal{X}}_\epsilon$ clearly has the proper form (1.80). It follows easily from (3.29) that the $\epsilon \to 0$ limit of

$$\mathcal{M}_{\mu(V_1' \cdots V_m')}\mathcal{T}_\epsilon(\underline{\lambda};\ V_1' \cdots V_m')$$

exists, so (W 3) implies the same for $\hat{\mathcal{X}}_\epsilon(V_1' \cdots V_m')$. The dependence only on the structure of $G(V_1' \cdots V_m')$ follows from (W 4).

Now we may define $\mathcal{X}_{\epsilon,r}'(\underline{\lambda};\ V_1' \cdots V_m')$, $\overline{\mathcal{R}}_{\epsilon,r}'(\underline{\lambda};\ V_1' \cdots V_m')$, and $\mathcal{R}_{\epsilon,r}'(\underline{\lambda};\ V_1' \cdots V_m')$ using (3.40) as finite renormalization, according to Definition 1.20. Our main theorem is then

*Theorem 3.21:* Let $\mathcal{R}_{\epsilon,r}'(\underline{\lambda};\ V_1 \cdots V_n)$ be defined as above. Then

$$(3.41) \qquad \mathcal{W}[\mathcal{T}_\epsilon(\underline{\lambda};\ V_1 \cdots V_n)] = \lim_{r \to 0+} \mathcal{R}_{\epsilon,r}'(\underline{\lambda}_0;\ V_1 \cdots V_n),$$

where $\underline{\lambda}_0 = (1, 1, \ldots, 1)$. We remark that the $\epsilon \to 0$ limit in (3.41), which we know exists, is the claimed equivalence of analytic renormalization with Definition 1.19.

We need one more preliminary result.

*Lemma 3.22:* The limits

$$\mathcal{X}_\epsilon'(\underline{\lambda};\ V_1' \cdots V_m') = \lim_{r \to 0+} \mathcal{X}_{\epsilon,r}'(\underline{\lambda};\ V_1' \cdots V_m'),$$

$$\overline{\mathcal{R}}_\epsilon'(\underline{\lambda};\ V_1' \cdots V_m') = \lim_{r \to 0+} \overline{\mathcal{R}}_{\epsilon,r}'(\underline{\lambda};\ V_1' \cdots V_m'),$$

$$\mathcal{R}_\epsilon'(\underline{\lambda};\ V_1' \cdots V_m') = \lim_{r \to 0+} \mathcal{R}_{\epsilon,r}'(\underline{\lambda};\ V_1' \cdots V_m'),$$

all exist for sufficiently large $\operatorname{Re}\lambda_\ell$, and the left hand sides may be analytically continued to lie in $\mathcal{B}(L, m)$.

*Proof:* Again we omit details. The quantities $\mathcal{X}_{\epsilon,r}'$, $\overline{\mathcal{R}}_{\epsilon,r}'$ and $\mathcal{R}_{\epsilon,r}$ may be calculated to give expressions similar to (3.2), with the lower limits on the integrals replaced by r.

The argument then proceeds as for $\mathcal{T}_\epsilon$. We note in particular that $\mathcal{X}'_\epsilon(\underline{\lambda};\ V'_1, \ldots, V'_m)$ has the form

$$\delta\Big(\sum_1^m p'_j\Big) \sum_{|\underline{i}| \leq \mu(V'_1 \cdots V'_m)} f_{\underline{i},\epsilon}(\underline{\lambda})(p')^{\underline{i}}\ , \tag{3.42}$$

where $\underline{i} = (i_{10}, i_{11}, \ldots, i_{m3})$, $|\underline{i}| = \Sigma_{j=1}^m\ \Sigma_{\mu=0}^3\ i_{j\mu}$, $(p')^{\underline{i}} = \Pi_{j=1}^m\ \Pi_{\mu=0}^4 (p_j{}^\mu)^{i_{j\mu}}$, and $[J_L(\underline{\lambda})]^{-1} f_{\underline{i},\epsilon}(\underline{\lambda})$ is an entire function of $\underline{\lambda}$.

*Proof of Theorem 3.21:* We first show that for $m' > 1$,

$$\mathcal{W}\mathcal{X}'_\epsilon(\underline{\lambda};\ V'_1, \ldots, V'_{m'}) = 0\ . \tag{3.43}$$

The statement is true (vacuously) for $m' = 1$; we assume it for all $1 \leq m' < m$, and consider an IPI graph $G(V'_1 \cdots V'_m)$. From (1.81) and (1.82),

$$\begin{aligned}\mathcal{X}'_{\epsilon,r}(\underline{\lambda};\ V'_1 \cdots V'_m) = -\mathfrak{M}_{\mu(V'_1 \cdots V'_m)} \Big\{ \sum_P \prod_{j=1}^{k(P)} \mathcal{X}'_{\epsilon,r}(\underline{\lambda};\ V^P_{j1} \cdots V^P_{jr(j)}) \\ \times \prod_{\text{conn}} \Delta^{(\ell)}_{\epsilon,r} \Big\} + \hat{\mathcal{X}}_\epsilon(V'_1 \cdots V'_m)\ .\end{aligned} \tag{3.44}$$

Consider a term from $\Sigma_P$ in (3.44) for which $r(j) > 1$ for some $j$, say $j = 1$ (note $k(P) \geq 2$, so $r(j) < m$ for all $j$). From (3.42) this term has the form

$$S_P(\underline{\lambda}, \epsilon, r) = \sum_{\underline{i}} f_{\underline{i},\epsilon,r}(\underline{\lambda}) \Big\{ \mathcal{F}^{-1}[\delta(\Sigma p')p'^{\underline{i}}] \prod_{j=2}^{k(P)} \mathcal{X}'_{\epsilon,r} \prod_{\text{conn}} \Delta \Big\}\ . \tag{3.45}$$

For sufficiently large $\operatorname{Re} \lambda_\ell$, we can let $r \to 0+$ in (3.45). The bracketed term converges to an element in $\mathcal{B}(L, m)$, and $f_{\underline{i},\epsilon,r}(\underline{\lambda})$ converges to $f_{\underline{i},\epsilon}(\underline{\lambda})$ [see (3.42)]. Actually, however $f_{\underline{i},\epsilon}(\underline{\lambda})$ depends only on those $\lambda_\ell$ such that both $V_{i_\ell}$ and $V_{f_\ell}$ lie in $\{V^P_{1,1}, \ldots, V^P_{1,r(1)}\}$, while the bracket in (3.45) depends on those $\lambda_\ell$ such that either $V_{i_\ell}$ or $V_{f_\ell}$ is not in this set. Thus (W 6) implies

$$\mathcal{W}[\lim_{r\to 0+} S_P(\underline{\lambda}, \epsilon, r)] = \sum_{\underline{i}} [\mathcal{W} f_{\underline{i},\epsilon}(\underline{\lambda})][\mathcal{W} \lim_{r\to 0+} \{\quad\}]\ . \tag{3.46}$$

But by the induction assumption,

$$\mathcal{W}\mathcal{X}'_\epsilon(\underline{\lambda};\ V^P_{1,1} \cdots V^P_{1,r(1)}) = 0\ ,$$

so that $\mathcal{W} f_{\underline{i},\epsilon} = 0$ and hence, from (3.46),

$$\mathcal{W}[\lim_{r\to 0} S_P(\underline{\lambda}, \epsilon, r)] = 0\ . \tag{3.47}$$

Now using Lemma 3.19,

$$(3.48)\qquad \mathcal{W}\mathfrak{X}'_\epsilon(\underline{\lambda};\ V'_1\cdots V'_m] = -\mathfrak{M}_{\mu(V'_1\cdots V'_m)}\,\mathcal{W}\{\sum_P [\lim_{r\to 0} S_P(\underline{\lambda},\epsilon,r]\}$$

$$+\hat{\mathfrak{X}}_\epsilon(V'_1\cdots V'_m)\ ,$$

since (W 2) implies $\mathcal{W}^2 = \mathcal{W}$. But by (3.47), all terms in $\Sigma_P$ in (3.48) vanish except for that partition P for which $k(P) = m$ (and hence $r(j) = 1$ for all $j$). But this term is exactly cancelled by $\hat{\mathfrak{X}}_\epsilon(V'_1\cdots V'_m)$; this proves (3.43).

Now (1.83) may be written

$$(3.49)\qquad \mathcal{R}'_{\epsilon,r}(\underline{\lambda};\ V_1\cdots V_n) = \prod_{\mathfrak{L}} \Delta^{(\ell)}_{\epsilon,r}(\lambda_\ell) + \sum_P{}' \prod_{j=1}^{k(P)} \mathfrak{X}'_{\epsilon,r}(\underline{\lambda};\ V^P_{j,1}\cdots V^P_{j,r(j)}) \prod_{\text{conn}} \Delta^{(\ell)}_{\epsilon,r}(\lambda_\ell),$$

where $\Sigma'_P$ is over all partitions P of $\{V_1,\ldots,V_n\}$ with $1 \le k(P) < n$. We let $r\to 0$ in (3.49) (for sufficiently large $\mathrm{Re}\,\lambda_\ell$), then apply $\mathcal{W}$ to both sides. Equation (3.43) and another use of (W 6) imply that $\mathcal{W}$ annihilates the second term on the right hand side. But the first term on this side is just $\mathcal{T}_\epsilon(\underline{\lambda};\ V_1,\ldots,V_n)$, by Lemma 3.16. Thus (3.49) becomes

$$\mathcal{W}\mathcal{R}'_\epsilon(\underline{\lambda};\ V_1\cdots V_n) = \mathcal{W}\mathcal{T}_\epsilon(\underline{\lambda}\,;\ V_1\cdots V_n)\ .$$

But now we use Theorem 1.21, which implies that $\mathcal{R}_\epsilon(\underline{\lambda};\ V_1,\ldots,V_n)$ is analytic at $\underline{\lambda}^0 = (1,1,\ldots,1)$, and hence by (W 2),

$$\mathcal{W}\mathcal{R}'_\epsilon(\underline{\lambda};\ V_1\cdots V_n) = \mathcal{R}'_\epsilon(\underline{\lambda}^0;\ V_1\cdots V_n)\ .$$

This completes the proof of Theorem 3.21.

# CHAPTER IV

## *Summation of Feynman Amplitudes*

### Section 1. INTRODUCTION.

In Chapter II we defined a generalized Feynman amplitude $\mathcal{T}$, depending on parameters $\underline{\lambda}$, $\underline{Q}$, and $\underline{e}$, which was (formally) equal to the Feynman amplitude of various graphs when the parameters took on certain discrete values. In Chapter III we exploited the $\underline{\lambda}$-dependence of $\mathcal{T}$ to circumvent the divergence difficulties associated with Feynman integrals and thus to define renormalized Feynman amplitudes in a new way. Here we turn to the properties and applications of the dependence of $\mathcal{T}$ on $\underline{Q}$ and $\underline{e}$.

Now if $G$ is a Feynman graph, $\underline{e}(G)$ its incidence matrix, and $Q^{(\ell)}_{rs}(G) = -e^{(\ell)}_r(G)e^{(\ell)}_s(G)$, the Feynman amplitude for $G$ is given formally by $\mathcal{T}[\underline{\lambda}^0, \underline{Q}(G), \underline{e}(G)]$ [with $\underline{\lambda}^0 = (1, 1, \ldots, 1)$]. Thus the GFA with general $\underline{Q}$ and $\underline{e}$ provides an interpolation between the Feynman amplitudes for different graphs. In Section 2 we discuss the continuity properties of this interpolation. Unfortunately, despite the natural way in which the interpolation was introduced, it is not necessarily continuous in the variables $\underline{Q}$ and $\underline{e}$. Continuity (or differentiability) is obtained only when $\mathrm{Re}\,\lambda_\ell$ is sufficiently large. This leads to difficulty in the applications.

Now let us consider the graphs which occur in the perturbation expansion of some truncated covariant time-ordered vacuum expectation value in a field theory. It was stated in Chapter II, Section 1, that a single GFA $\mathcal{T}(\underline{\lambda}, \underline{Q}, \underline{e})$ gives the amplitudes for many of the graphs $G$ in the expansion if we take $\underline{Q} = \underline{Q}(G)$, $\underline{e} = \underline{e}(G)$. (Actually, something must be said about the coefficient in equation (2.2), which was not included in the GFA; we discuss this in Section 3.) Let $\{G_1, \ldots, G_P\}$ be this set of graphs; we are interested in the quantity

$$\sum_{p=1}^{P} \mathcal{T}(\underline{\lambda}^0, \underline{Q}(G_p), \underline{e}(G_p)) \,. \tag{4.1}$$

The original motivation for the introduction of the parameters $\underline{e}$ and $\underline{Q}$ was the hope that the sum (4.1) could be converted into an integral (over the variables $\underline{Q}$ and $\underline{e}$) involving the GFA. That is, we would like to find an integral $I(\underline{\lambda})$ such that

$$I(\underline{\lambda}) = \sum_{p=1}^{P} \mathcal{T}(\underline{\lambda}, \underline{Q}(G_p), \underline{e}(G_p)] \,. \tag{4.2}$$

Note that, according to Chapter III, the sum of the renormalized Feynman amplitudes for $G_1, \ldots, G_P$ would be given by $\mathcal{E}I$, where $\mathcal{E}$ is a generalized evaluator. Such a representation might, for example, display explicitly some of the cancellations which occur when amplitudes involving fermions are summed, and thus contribute to a discussion of the convergence (or divergence) of the perturbation series. It might also be used to study the analytic behavior (in the momenta) of the sum (4.1).

We have been only partially successful in our search for such a representation, as is discussed in Section 4. Specifically, an integral representation $I(\underline{\lambda})$ as in (4.2) has been found which is valid for sufficiently large $\operatorname{Re}\lambda_\ell$; its failure for other values is connected with the non-continuity of $\mathcal{T}(\underline{\lambda}, Q, \underline{e})$ discussed above. Because each individual GFA on the right hand side of (4.2) is meromorphic in $\underline{\lambda}$, the integral $I(\underline{\lambda})$ must be also. However, what we lack is an explicit analytic continuation of $I(\underline{\lambda})$ itself to a neighborhood of the physically interesting point $\lambda_\ell = 1$. No applications of the method seem to be possible until such a continuation is found.

## Section 2. Q DEPENDENCE OF THE GFA.

Before discussing the dependence of $\mathcal{T}$ on $Q$ and $\underline{e}$, we give a theorem about its $\underline{\lambda}$ dependence.

*Theorem 4.1:* For fixed $Q$ and $\underline{e}$, the GFA $\mathcal{T}(\underline{\lambda}, Q, \underline{e})$ [see equation (2.34)] has a meromorphic extension to all $\underline{\lambda} \in \mathbb{C}^L$. If we denote the extension also by $\mathcal{T}(\underline{\lambda}, Q, \underline{e})$, then

$$\prod_K \left\{ \Gamma\left[ \sum_{\ell \in K} (\lambda_\ell - r_\ell - 2) \right] \right\}^{-1} \mathcal{T}(\underline{\lambda}, Q, \underline{e})$$

is analytic on $\mathbb{C}^L$, where $\Pi_K$ runs over all non-empty subsets $K \subset \{1, \ldots, L\}$.

*Proof:* We will not give details of the proof. It is almost exactly like the proof of Theorem 3.4, except that the variable change (2.30) is used instead of (3.10), and Lemmas 3.6 and 3.7 are replaced by an argument similar to that used in the proof of Lemma 2.13.

Let us now discuss more completely the set of L-tuples of quadratic forms $Q$ which may occur as arguments of the GFA $\mathcal{T}(\underline{\lambda}, Q, \underline{e})$. Recall first that each $Q^{(\ell)}$ was symmetric and translation invariant. We let F denote the set of all such $Q$, and identify F with $R^{\frac{1}{2}n(n-1)L}$, specifying $Q \in F$ by the $[\frac{1}{2}n(n-1)L]$- tuple of real numbers $\{Q^{(\ell)}_{rs} \mid 1 \le s < r \le n, 1 \le \ell \le L\}$. Let $F^+ \subset F$ be the closed subset $\{Q \mid Q^{(\ell)}_{rs} \ge 0$ for all $1 \le s < r \le n,\ 1 \le \ell \le L\}$, and let $O \subset F$ be the open subset $\{Q \mid \Sigma_{\ell=1}^{L} Q^{(\ell)}$ is positive definite on $E\}$. Then the requirement of Chapter II, Section 5, that $Q$ satisfy conditions (Q1) and (Q2), becomes precisely $Q \in F^+ \cap O$.

*Remark 4.2:* For notational convenience we let $\Gamma_0 = \{(\ell, r, s) \mid 1 \le \ell \le L,\ 1 \le s < r \le n\}$. We will also refer to the graph $G_0$ which has vertices $V_1, \ldots, V_n$ and lines indexed by $\Gamma_0$; specifically, if $\gamma = (\ell, r, s) \in \Gamma_0$, the line $\gamma$ of $G_0$ has initial vertex $V_s$ and final vertex $V_r$. Given $Q \in F^+ \cap O$ and an L-tuple of positive real numbers $a_1, \ldots, a_L$, we associate with $\gamma = (\ell, r, s) \in \Gamma_0$ the inverse Feynman parameter

$$\beta_\gamma = \frac{Q^{(\ell)}_{rs}}{a_\ell}.$$

*Definition 4.3:* Let $f(\underline{Q})$ be a complex function defined on $F^+ \cap O$. We say that $f \in C^0$ if $f$ is continuous; $f \in C^1$ if all derivatives of $f$ with respect to $Q_{rs}^{(\ell)}$ $(r > s)$, including one-sided derivatives at boundaries, exist and are continuous in $F^+ \cap O$. (That is, each limit

$$\left[\frac{\partial}{\partial Q_{rs}^{(\ell)}} f\right](\underline{Q}) = \lim_{\epsilon \to 0} \frac{f(\underline{Q}') - f(\underline{Q})}{\epsilon} ,$$

with $Q'^{(\ell_1)}_{r_1 s_1} = Q^{\ell_1}_{r_1 s_1} + \delta_{rr_1} \delta_{ss_1} \delta_{\ell\ell_1} \epsilon$ and $\underline{Q} \in F^+ \cap O$, $\underline{Q}' \in F^+ \cap O$, exists and is continuous.) Inductively, we say that $f \in C^n$ if $f \in C^{n-1}$ and each $(n-1)^{st}$ derivative of $f$ is in $C^1$.

*Remark 4.4:* This is not the definition usually given of the class $C^n$ for a non-open set (it appears weaker), but it is in fact equivalent. That is, if $f \in C^n$ in the sense of Definition 4.3, there is an extension $f'$ of $f$ to some open set $U$ with $F^+ \cap O \subset U$, such that $f'$ is $C^n$ on $U$.

*Proof:* The proof depends, of course, on the special nature of the set $F^+ \cap O$. We will construct the extension $f'$ and leave the verification of its properties to the reader. Let $\underline{P}$ be a boundary point of $F^+ \cap O$, so that

$$P_{rs}^{(\ell)} = 0 \qquad \text{if } (\ell, r, s) \in \Gamma$$

$$P_{rs}^{(\ell)} > 0 \qquad \text{if } (\ell, r, s) \in \Gamma'$$

with $\Gamma \subset \Gamma_0$ nonempty. We define $f'$ at all points $\underline{Q} \in F$ satisfying

$$\text{(4.4)} \qquad \begin{aligned} Q_{rs}^{(\ell)} &\le 0 && \text{if } (\ell, r, s) \in \Gamma \\ Q_{rs}^{(\ell)} &= P_{rs}^{(\ell)} && \text{if } (\ell, r, s) \in \Gamma' \end{aligned}$$

by the Taylor series of $f$ at $\underline{P}$ in the variables $Q_{rs}^{(\ell)}$, with $(\ell, r, s) \in \Gamma$, up to order $n$. That is,

$$\text{(4.5)} \qquad f'(\underline{Q}) = \sum_{i=0}^{n} \frac{1}{i!} \sum_{\substack{(\ell_j, r_j, s_j) \in \Gamma \\ j = 1, \ldots, i}} \left[\frac{\partial}{\partial Q^{(\ell_1)}_{r_1 s_1}} \cdots \frac{\partial}{\partial Q^{(\ell_i)}_{r_i s_i}} f\right](\underline{P})\, Q^{(\ell_1)}_{r_1 s_1} \cdots Q^{(\ell_i)}_{r_i s_i} ,$$

where the derivatives in (4.5) are the one-sided derivatives of Definition 4.3. It may be verified that the union of $F^+ \cap O$ with the set of all $\underline{Q}$ satisfying (4.4) for some $\underline{P}$ is an open neighborhood of $F^+ \cap O$, and that $f'$ is in $C^n$ there.

We can now state the principal theorem of this section.

*Theorem 4.5:* Let $\mathcal{J}(\underline{\lambda}, \underline{Q}, \underline{e})$ be the generalized Feynman amplitude (2.34) [or (2.42)], with $\underline{Q} \in F^+ \cap O$. Then $\mathcal{J} \in C^m$ when $\underline{\lambda}$ satisfies

$$\operatorname{Re} \lambda_\ell > 2 + m + r_\ell .$$

(Note in particular that the GFA, originally defined for all $\operatorname{Re}\lambda_\ell > 2 + r_\ell$, is continuous in Q there.) The Q-derivatives may be interchanged with the $\alpha$-integrations.

We need two lemmas; in stating them, we write $A = A(G_0)$ (see Definition A.8 and Remark 4.2).

*Lemma 4.6:* Let $U_1, \ldots, U_k$, $W_1, \ldots, W_k$ be subsets of $\{1, \ldots, n\}$, let $r_1, \ldots, r_m$, $s_1, \ldots, s_m$ satisfy $1 \le s_i < r_i \le n$ for all i. For notational convenience we write $U_{k+i} = \{r_i\}$, $W_{k+i} = \{s_i\}$ for $i = 1, \ldots, m$, and if $X = \{i_1, \ldots, i_j\}$ is a subset of $\{1, 2, \ldots, k, k+1, \ldots, k+m\}$, write $B(X) = A(U_{i_1} | W_{i_1}, \ldots, U_{i_j} | W_{i_j})$ (see Definition A.14). Then

(a) $$\frac{\partial}{\partial Q^{(\ell)}_{r_1 s_1}} A(\tfrac{i}{i}) = \frac{1}{\alpha_\ell} A(r_1 | s_1);$$

(b) $$\frac{\partial}{\partial Q^{(\ell)}_{r_1 s_1}} A(U_1 | W_1, \ldots, U_k | W_k) = \frac{1}{\alpha_\ell} A(U_1 | W_1, \ldots, U_k | W_k, r_1 | s_1);$$

(c) $$\frac{\partial}{\partial Q^{(\ell_1)}_{r_1 s_1}} \cdots \frac{\partial}{\partial Q^{(\ell_m)}_{r_m s_m}} A(\tfrac{i}{i})^{-q}$$

$$= \frac{1}{\alpha_{\ell_1} \cdots \alpha_{\ell_m}} \sum_P \frac{(-1)^{k(P)}[k(P)+q-1]!}{(q-1)!\, A(\tfrac{i}{i})^{q+k(P)}} \prod_{i=1}^{k(P)} B(X_i)$$

where $\Sigma_P$ is over all partitions of $\{k+1, \ldots, k+m\}$ into $k(P)$ sets $X_1, \ldots, X_{k(P)}$;

(d) $$\frac{\partial}{\partial Q^{(\ell_1)}_{r_1 s_1}} \cdots \frac{\partial}{\partial Q^{(\ell_m)}_{r_m s_m}} \left[\frac{A(U_1 | W_1, \ldots, U_k | W_k)}{A(\tfrac{i}{i})^q}\right]$$

$$= \frac{1}{\alpha_{\ell_1} \cdots \alpha_{\ell_m}} \sum_P{}' \frac{(-1)^{[k(P)+1]}[k(P)+q-2]!}{(q-1)!\, A(\tfrac{i}{i})^{[q+k(P)-1]}} \prod_{i=1}^{k(P)} B(X_i)$$

where $\Sigma'_P$ is over all partitions of $\{1, \ldots, k+m\}$ into $k(P)$ sets $X_1, \ldots, X_{k(P)}$, with $\{1, \ldots, k\} \subset X_1$.

*Proof:* (a) and (b) follow directly from the Definitions and Lemma A.9; (c) and (d) are easily proved by induction on m, using (a) and (b).

*Lemma 4.7:* Let $U_1, \ldots, U_k$, $W_1, \ldots, W_k$ be subsets of $\{1, \ldots, n\}$. Then the quantities

(4.7) $$A(\tfrac{i}{i})^{-1}$$

and

(4.8) $$A(U_1 | W_1, \ldots, U_k | W_k)\, A(\tfrac{i}{i})^{-1}$$

are uniformly bounded when $Q$ and $\underline{\alpha}$ are restricted to compact subsets of $F^+ \cap O$ and $\{\underline{\alpha} \mid \alpha_\ell \geq 0\}$ respectively.

*Proof:* We give the proof for (4.8); the proof for (4.7) is similar. Suppose that the compact subset of $\underline{\alpha}$-space has the form $0 \leq \alpha_\ell \leq M$. From Definition A.14 and Lemma A.9,

$$\frac{A(U_1|W_1, \ldots, U_k|W_k)}{A(\tfrac{i}{i})} = \frac{\Sigma_{T_{k+1}} \prod_{\gamma \in T_{k+1}} \frac{Q_{rs}^{(\ell)}}{\alpha_\ell}}{\Sigma_T \prod_{\gamma \in T} \frac{Q_{rs}^{(\ell)}}{\alpha_\ell}} , \tag{4.9}$$

where the sum in the denominator is over all trees $T$ of $G_0$, and we have taken $\gamma = (\ell, r, s)$. For any fixed $(k+1)$-tree $T_{k+1}$ from the numerator of (4.9) we have

$$A(\tfrac{i}{i}) \geq \sum_{T \supset T_{k+1}} \prod_{\gamma \in T} \frac{Q_{rs}^{(\ell)}}{\alpha_\ell} ,$$

since $Q_{rs}^{(\ell)}$ and $\alpha_\ell$ are non-negative. Thus

$$\begin{aligned} A(\tfrac{i}{i})^{-1} \left[ \prod_{\gamma \in T_{k+1}} \frac{Q_{rs}^{(\ell)}}{\alpha_\ell} \right] &\leq \left[ \sum_{T \supset T_{k+1}} \prod_{\gamma \in (T - T_{k+1})} Q_{rs}^{(\ell)}/\alpha_\ell \right]^{-1} \\ &\leq M^k \left[ \sum_{T \supset T_{k+1}} \prod_{\gamma \in (T - T_{k+1})} Q_{rs}^{(\ell)} \right]^{-1} . \end{aligned} \tag{4.10}$$

But for $Q \in F^+ \cap O$ the subgraph of $G_0$ formed by those lines $\gamma = (\ell, r, s)$ with $Q_{rs}^{(\ell)} > 0$ is connected (see Chapter II, Section 5). This implies that for any $(k+1)$-tree $T_{k+1}$ there is some tree $T \supset T_{k+1}$ with $Q_{rs}^{(\ell)} > 0$ for every $\gamma = (\ell, r, s) \in (T - T_{k+1})$. Thus the bracket in (4.10) cannot vanish on $F^+ \cap O$; this completes the proof.

*Proof of Theorem 4.5:* From (2.34), $\mathcal{T}$ has the form

$$\mathcal{T}(\underline{\lambda}, Q, \underline{e}) = \lim_{\epsilon \to 0} \int_0^\infty \cdots \int_0^\infty \prod_1^L \alpha_\ell^{\lambda_\ell - 3 - r_\ell} d\alpha_\ell f(\underline{\alpha}, Q, \underline{e}) e^{-i\epsilon \Sigma_1^L \alpha_\ell} .$$

The variable $Q$ appears in $f$ linearly and through the quantities $A(\tfrac{i}{i})^{-2}$ and $A_E^{-1}$; thus repeated $Q$-derivatives of $f$ may be calculated using Lemma 4.6(b), (c). Lemma 4.7 then implies that, for $Q$ in a compact subset of $F^+ \cap O$ and $\alpha_\ell \geq 0$, the quantity

$$\alpha_{\ell_1} \cdots \alpha_{\ell_m} \frac{\partial}{\partial Q_{r_1 s_1}^{(\ell_1)}} \cdots \frac{\partial}{\partial Q_{r_m s_m}^{(\ell_m)}} f(\underline{\alpha}, Q, \underline{e})$$

is bounded by a polynomial in $\underline{\alpha}$. Since when $\underline{\lambda}$ satisfies (4.6) the quantity

$$\left[ \prod_1^L \alpha_\ell^{\lambda_\ell - r_\ell - 3} \right] \left[ \alpha_{\ell_1} \cdots \alpha_{\ell_m} \right]^{-1} e^{-i\epsilon \Sigma_1^L \alpha_\ell}$$

is integrable over the region $\alpha_\ell \geq 0$, the theorem follows easily.

*Remark 4.8:* (a). It is easily seen that theorem 4.1 applies with only slight modification to any $\underline{Q}$-derivative of $\mathcal{T}$.
(b). It follows immediately from (2.34) that $\mathcal{T}(\underline{\lambda}, \underline{Q}, \underline{e})$ is $C^\infty$ in the variables $\underline{e}$.

Section 3. GENERALIZED FEYNMAN AMPLITUDES IN A FIELD THEORY

In this section we consider a field theory of the type discussed in Chapter I and study the relations between generalized Feynman amplitudes and the Feynman amplitudes occurring in the perturbation series for some truncated covariant TOVEV [equation 1.60)]. The GFA $\mathcal{T}$ has been defined to depend implicitly on positive integers L and n and explicitly on L-tuples $\underline{\lambda}$, $\underline{Q}$, $\underline{e}$, $\underline{Z}$, $\underline{m}$; $\mathcal{T}$ becomes formally equal to the Feynman amplitude of various graphs for certain values of these variables. Such a graph always has n vertices and L lines; the variables $\underline{Z}$ and $\underline{m}$ determine the propagators associated with the lines, and $\underline{Q}$ and $\underline{e}$ determine the arrangement of lines in the graph.

Now the amplitude for the graph G in the series (1.60) has the form

$$A(G)_{\alpha_1\cdots\alpha_R}(x_1\cdots x_R) = \frac{(-ig)^m}{m!} \sum_{\{\alpha_v|\rho(v)>R\}} \prod_{j=R+1}^{R+m} M^{(q(j))}_{\alpha_{j,1}\cdots\alpha_{j,s(q(j))}}$$

(4.11)

$$\int\cdots\int dx_{R+1}\cdots dx_{R+m}\,\sigma \prod_{(v_1,v_2)\in\mathcal{L}} \overline{\psi^{(\pi(v_1))}_{\alpha_{v_1}}(x_{\rho(v_1)})\,\psi^{(\pi(v_2))}_{\alpha_{v_2}}}(x_{\rho(v_2)})\ .$$

[This is equation (1.59); the notation is explained in the discussion preceding that equation.] For $\ell = (v_1, v_2) \in \mathcal{L}$, the propagator

$$\Delta^{(\ell)}_{\alpha\beta}(x-y) = \overline{\psi^{(\pi(v_1))}_{\beta}(y)\,\psi^{(\pi(v_2))}_{\alpha}}(x)$$

depends only on the type of the field $\psi^{(\pi(v_1))}$. Thus if we consider all amplitudes (4.11) for some fixed truncated TOVEV, with a fixed order m and a fixed map $q: \{R+1, \ldots, R+m\} \to \Theta$, the set of propagators occurring in the product

$$\prod_{\ell=(v_1,v_2)\in\mathcal{L}} \Delta^{(\ell)} \tag{4.12}$$

is independent of the particular graph involved (i.e., of the choice of $\mathcal{L} \subset \mathcal{V}^q \times \mathcal{V}^q$). All products (4.12) for this set of amplitudes may thus be obtained (formally) as values of the GFA $\mathcal{T}(\underline{\lambda}, \underline{Q}, \underline{e}, \underline{Z}, \underline{m})$ by varying $\underline{Q}$ and $\underline{e}$ while keeping $\underline{Z}$ and $\underline{m}$ fixed.

Let us state this more precisely. The basic fields of our theory are $\Phi^{(1)}, \ldots, \Phi^{(I)}$; we will say that a line $\ell$ of a graph is of type i if it arises from the contraction of $\phi^{(i)}$ with itself (for $\phi^{(i)}$ SCC) or with its adjoint ($\phi^{(i)}$ non-SCC). The fields $\xi^{(i)}$ were defined in Definition 1.1; we write

$$\overline{\xi^{(i)}_{\beta}(y)\,\xi^{(i)}_{\alpha}}(x) = Y^{(i)}_{\alpha\beta}\left(\frac{\partial}{\partial x}\right)\Delta_F(x-y;\,M_i)\,,$$

$$\overline{\xi^{(i)*}_{\beta}(y)\,\xi^{(i)}_{\alpha}}(x) = Y^{(i)}_{\alpha\beta}\left(\frac{\partial}{\partial x}\right)\Delta_F(x-y;\,M_i)\,,$$

for $\phi^{(i)}$ SCC or non-SCC respectively, where $M_i$ is the mass of the particles associated with $\Phi^{(i)}$. Now recall the definition of $\mathcal{G}(\Theta, \delta, m, q)$ given in Chapter I, Section 3 (D, iii,d). Each $G \in \mathcal{G}(\Theta, \delta, m, q)$ has the same number $L_i$ of lines of type i (this number is determined by $\mathcal{O}^q$). Take $L = \Sigma_1^I L_i$, and set

$$\left.\begin{aligned} Z^{(\ell)}_{\alpha_\ell\beta_\ell} &= Y^{(i)}_{\alpha_\ell\beta_\ell} \\ m_\ell &= M_i \end{aligned}\right\} \quad \text{whenever } \sum_1^{i-1} L_j < \ell \le \sum_1^{i} L_i\,,$$

$$\mathcal{T}_{\alpha_1\beta_1\cdots\alpha_L\beta_L}(\underline{\lambda}, \underline{Q}, \underline{e}) = \mathcal{T}(\underline{\lambda}, \underline{Q}, \underline{e}, \underline{Z}, \underline{m})\,.$$

Then if $G \in \mathcal{G}(\Theta, \delta, m, q)$, and we number the lines of G so that lines $1,\dots,L_1$ are of type 1, etc., and orient charged lines so that their initial vertex is at $\phi^{(i)*}$, the amplitude (4.11) becomes

$$A(G)_{\alpha_1\cdots\alpha_R}(x_1\cdots x_R) = \frac{(-ig)^m}{m!}\sum_{\{\alpha_v|\rho(v)>R\}}\ \prod_{j=R+1}^{R+m} M^{(q(j))}_{\alpha_{j,1}\cdots\alpha_{j,s(q(j))}} \tag{4.13}$$
$$\times\int\cdots\int dx_{R+1}\cdots dx_{R+m}\,\sigma\,\mathcal{T}_{\rho_1,\sigma_1\cdots\rho_L,\sigma_L}(\underline{\lambda}^0, \underline{Q}(G), \underline{e}(G))\,.$$

Here $\underline{e}(G)$ is the incidence matrix, $Q^{(\ell)}_{rs}(G) = -e^{(\ell)}_r(G)e^{(\ell)}_s(G)$, $\underline{\lambda}^0 = (1, 1, \dots, 1)$, and, for $\ell = (v_1, v_2)$, $\rho_\ell = \alpha_{v_2}$, $\sigma_\ell = \alpha_{v_1}$.

Now our goal is to find an interpolating quantity $A(\underline{\lambda}, \underline{Q}, \underline{e})$ which would take the value (4.13) for $\underline{\lambda} = \underline{\lambda}^0$, $\underline{Q} = \underline{Q}(G)$, $\underline{e} = \underline{e}(G)$.. The hard part is already contained in (4.13) through the interpolating abilities of $\mathcal{T}$. But (4.13) still contains a dependence on the structure of the graph through $\sigma$ and the indices $\rho_\ell$, $\sigma_\ell$. For a general field theory such an interpolation may be defined in various ways; at the moment we have no basis for choosing one over the other. Instead of giving such a general construction, we will discuss a special case which illustrates the important features but has a relatively simple form.

Thus we consider the coupling of a Dirac field $\Psi_a$ $(a = 1, 2, 3, 4)$ with a massive SCC vector field $\Phi_\mu$ $(\mu = 0, 1, 2, 3)$ through the interaction

$$\mathcal{L}_I(x) = g:\ \bar{\Psi}(x)\gamma^\mu\Psi(x)\Phi_\mu(x)\ : \tag{4.14}$$

[We take a massive field $\Phi$ since we have assumed positive mass throughout our discussion, but the in combinatories to follow $\Phi$ could be replaced by the electromagnetic vector potential A; then (4.14) becomes the usual electromagnetic coupling.] Note that $\gamma^0$, $\gamma^1$, $\gamma^2$, $\gamma^3$ are the usual $4 \times 4$ Dirac matrices and $\bar{\Psi} = \Psi^\dagger \gamma^0$.

Let us study a typical truncated covariant TOVEV in the theory, say

$$(\Omega, P^*[\bar{\Psi}_{\alpha_1}(x_1)\Psi_{\beta_2}(x_2) \cdots \bar{\Psi}_{\alpha_{2p-1}}(x_{2p-1})\Psi_{\beta_{2p}}(x_{2p}) \times \Phi_{\mu_{2p+1}}(x_{2p+1}) \cdots \Phi_{\mu_{2p+q}}(x_{2p+q})]\Omega)^T . \tag{4.15}$$

Any graph G of order m in the expansion (1.60) of (4.15) has $L_1 = p+m$ fermion lines and $L_2 = \frac{1}{2}(q+m)$ boson lines; we write $\mathfrak{L}_1 = \{1, \ldots, L_1\}$, $\mathfrak{L}_2 = \{L_1+1, \ldots, L_1+L_2\}$, and $L = L_1+L_2$. Thus, defining

$$\overline{\bar{\psi}_\alpha(y)\psi_\beta}(x) = Y^{(1)}_{\beta\alpha}\left(\frac{\partial}{\partial x}\right) \Delta_F(x-y;\, M_1) ,$$

$$Z^{(\ell)}_{\beta\alpha} = Y^{(1)}_{\beta\alpha} ,$$

$$m_\ell = M_1 ,$$

for $\ell = 1, \ldots, L_1$, and

$$\overline{\phi_{\mu_2}(y)\,\phi_{\mu_1}}(x) = Y^{(2)}_{\mu_1\mu_2}\left(\frac{\partial}{\partial x}\right) \Delta_F(x-y,\, M_2) ,$$

$$Z^{(\ell)}_{\mu_1\mu_2} = Y^{(2)}_{\mu_1\mu_2} ,$$

$$m_\ell = M_2 ,$$

for $\ell = L_1+1, \ldots, L_1+L_2$, we have from (4.13)

$$A(G)_{\alpha_1,\beta_2,\ldots,\alpha_{2p-1},\beta_{2p},\mu_{2p+1}\cdots\mu_{2p+q}}(x_1, \ldots, x_{2p+q}) = \frac{(-ig)^m}{m!} \sum_{\{\alpha_i,\beta_i,\mu_i | 2p+q+1 \le i \le 2p+q+m\}} \prod_{i=2p+q+1}^{2p+q+m} \gamma^{\mu_i}_{\alpha_i\beta_i} \int dx_{2p+q+1} \cdots dx_{2p+q+m}\, \sigma\mathcal{T}_{\beta_{i_1}\alpha_{f_1}\cdots\beta_{i_{L_1}}\alpha_{f_{L_1}}\cdots\mu_{i_L}\mu_{f_L}}[\underline{\lambda}^0, \underline{Q}(G), \underline{e}(G)] . \tag{4.16}$$

Now define two more matrices associated with the graph G,

$$i^{(\ell)}_r(G) = \begin{cases} 1 & \text{if } i_\ell = r \\ 0 & \text{otherwise} , \end{cases}$$

$$f^{(\ell)}_r(G) = \begin{cases} 1 & \text{if } f_\ell = r \\ 0 & \text{otherwise} , \end{cases}$$

and note that

$$(4.17\,a) \qquad i_r^{(\ell)}(G) = -\tfrac{1}{2}\,[Q_{rr}^{(\ell)}(G) + e_r^{(\ell)}(G)]\,,$$

$$(4.17\,b) \qquad f_r^{(\ell)}(G) = \tfrac{1}{2}\,[e_r^{(\ell)}(G) - Q_{rr}^{(\ell)}(G)]\,.$$

Let $\mathcal{H} = \{1, 3, \ldots, 2p-1, 2p+q+1, \ldots, 2p+q+m\}$ be the set of initial vertices of fermion lines, $\mathcal{J} = \{2, 4, \ldots, 2p, 2p+q+1, \ldots, 2p+q+m\}$ their final vertices, and $\mathcal{K} = \{2p+1, \ldots, 2p+q+m\}$ the boson vertices, and let $\eta : \mathcal{J} \to \mathcal{H}$ be defined by

$$\eta(i) = \begin{cases} i-1, & \text{if } i \leq 2p \\ i\,, & \text{if } i > 2p\,. \end{cases}$$

Then the GFA in (4.16) may be written

$$(4.18) \qquad \begin{aligned} \mathcal{T}_{\beta_{i_1}\alpha_{i_1}\cdots\mu_{i_L}\mu_{f_L}}(\underline{\lambda}, Q, \underline{e}) &= \prod_{\ell=1}^{L_1} i_{h_\ell}^{(\ell)}(G)\, f_{j_\ell}^{(\ell)}(G) \prod_{\ell=L_1+1}^{L_1+L_2} i_{k_\ell}^{(\ell)}(G)\, f_{k'_\ell}^{(\ell)}(G) \\ &\times \mathcal{T}_{\beta_{h_1}\alpha_{j_1}\cdots\beta_{h_{L_1}}\alpha_{j_{L_1}}\mu_{k_{L_1+1}}\mu_{k'_{L_1+1}},\ldots,\mu_{k_L}\mu_{k'_L}}(\underline{\lambda}, Q, \underline{e})\,, \end{aligned}$$

where $h_\ell = i_\ell$, $j_\ell = f_\ell$, $k_\ell = i_\ell$, $k'_\ell = f_\ell$. We actually consider h, j, k, k′ as maps $h: \mathcal{L}_1 \to \mathcal{H}$, $j: \mathcal{L}_1 \to \mathcal{J}$, $k, k': \mathcal{L}_2 \to \mathcal{K}$; then it is easily verified that the sign factor $\sigma$ in (4.16) is given by the sign of the permutation $\eta j h^{-1}$ of $\mathcal{H}$:

$$\sigma = \pi(\eta j h^{-1})\,.$$

Now (4.18) seems to be a completely trivial modification of (4.16). Because i(G) and f(G) have only one non-zero entry for each $\ell$, however, we may insert in (4.18) a summation over *all* maps $h: \mathcal{L}_1 \to \mathcal{H}$, $j: \mathcal{L}_1 \to \mathcal{J}$, $k, k': \mathcal{L}_2 \to \mathcal{K}$, where h, j, k and k′ are all 1 – 1 and onto, without changing the value. Thus defining

$$(4.19) \qquad \begin{aligned} A(\underline{\lambda}, Q, \underline{e})_{\alpha_1,\beta_2,\ldots,\mu_{2p+q}}(x_1, \ldots, x_{2p+q}) &= \frac{(-ig)^m}{m!} \sum_{h,j,k,k'} \ \sum_{\{\alpha_i,\beta_i,\mu_i \,|\, 2p+q+1 \leq i \leq 2p+q+m\}} \\ &\prod_{i=2p+q+1}^{2p+q+m} \gamma^{\mu_i}_{\alpha_i\beta_i} \prod_{\ell=1}^{L_1} \left[ i_{h_\ell}^{(\ell)} f_{j_\ell}^{(\ell)} \right] \prod_{\ell=L_1+1}^{L_1+L_2} \left[ i_{k_\ell}^{(\ell)} f_{k'_\ell}^{(\ell)} \right] \pi(\eta j h^{-1}) \\ &= \mathcal{T}(\underline{\lambda}, Q, \underline{e})_{\beta_{h_1},\alpha_{j_1}\cdots\mu_{k_L}\mu_{k'_L}} \end{aligned}$$

where $\underline{i}$ and $\underline{f}$ are given by $i_r^{(\ell)} = -\tfrac{1}{2}(e_r^{(\ell)} + Q_{rr}^{(\ell)})$, $f_r^{(\ell)} = \tfrac{1}{2}(e_r^{(\ell)} - Q_{rr}^{(\ell)})$ [see (4.17)], we have the important relation

$$A[\underline{\lambda}^0, \underline{Q}(G), \underline{e}(G)] = A(G) \ . \tag{4.20}$$

Thus (4.19) is the desired interpolation.

*Remark 4.9:* (a) The amplitude $A(\underline{\lambda}, \underline{Q}, \underline{e})$ has the same properties of differentiability, etc., which were found for $\mathcal{I}$ in Section 2.

(b) Note that for an interaction Lagrangian such as (4.14) which contains only one term, *all* $m^{th}$ order graphs in the series for (4.15) are obtained from one amplitude (4.20), since $\mathcal{G}(\Theta, \delta, m, q) = \mathcal{G}(\Theta, \delta, m)$ for any $q$.

Section 4. SUMMATION OF AMPLITUDES.

We now turn to the problem, discussed in the introduction, of converting a sum of amplitudes into an integral. Our (partial) answer is based on

*Lemma 4.10:* Let $\sigma_n$ be the standard $n-1$ simplex in $R^n$, i.e., $\sigma_n = \{x \mid \Sigma_1^n x_i = 1, x_i \geq 0\}$, and let $\{v_i \mid i = 1, \ldots, n\}$ be the vertices of $\sigma_n$. Let $\mu$ be Lebesgue measure on $\sigma_n$ normalized to a total area $1/(n-1)!$. Then there is a polynomial of degree $n-1$,

$$P_n\left(\frac{\partial}{\partial x_1}, \ldots, \frac{\partial}{\partial x_n}, x_1 \cdots x_n\right) ,$$

such that if $f$ is any $C^{n-1}$ function on $\sigma_n$,

$$\sum_{i=1}^{n} f(v_i) = \int_{\sigma_n} [P_n f] \, d\mu \ . \tag{4.21}$$

*Proof:* We will use the notation of differential forms [9]. Thus let $\sigma_n$ be oriented in the usual way, so that

$$d\mu = (-)^i \, dx_1 \cdots \widehat{dx_i} \cdots dx_n \tag{4.22}$$

for any $1 \leq i \leq n$. Let $\sigma_{n,k}$ be the face of $\sigma_n$ opposite $v_k$, with the inherited orientation, and define

$$D_{ij} = \begin{cases} dx_1 \cdots \widehat{dx_i} \cdots \widehat{dx_j} \cdots dx_n \ , & \text{if } i < j \ ; \\ -D_{ji} \ , & \text{if } i > j \ . \end{cases}$$

The proof proceeds by induction on $n$. For $n = 2$ we take

$$P_2 = \partial_1 x_1 + \partial_2 x_2 \ .$$

Suppose then that we have defined $P_{n-1}$. Set

$$P_{n-1}^{(k)} = P_{n-1}(\partial_1 \cdots \hat{\partial}_k \cdots \partial_n, \ x_1, \ldots, \hat{x}_k, \ldots, x_n) \ ;$$

$$Q^{ij} = x_i P_{n-1}^{(j)} - x_j P_{n-1}^{(i)} \ .$$

Then we may take

$$P_n = -\frac{1}{(n-1)} \sum_{i \neq j} \partial_i Q^{ij} .$$

To see this, we first define the differential form

$$Q = \sum_{i<j} (-)^{i+j} [Q^{ij} f] D_{ij} .$$

We calculate easily, using (4.22),

$$dQ = -(n-1)(P_n f) dx_2 \cdots dx_n .$$

So by Stokes' theorem

$$\begin{aligned} \int_{\sigma_n} (P_n f) d\mu &= -\frac{1}{n-1} \sum_{k=1}^{n} (-)^k \int_{\sigma_{n,k}} Q \\ &= -\frac{1}{(n-1)} \sum_{k,i=1}^{n} (-)^i \int_{\sigma_{n,k}} [Q^{ik} f] D_{ik} \\ &= \frac{1}{(n-1)} \sum_{i,k=1}^{n} \int_{\sigma_{n,k}} [Q^{ik} f] \begin{cases} dx_2 \cdots \widehat{dx_k} \cdots dx_n & (k > 1) \\ dx_3 \cdot \cdot \cdot dx_n & (k = 1) . \end{cases} \end{aligned}$$

But $x_k = 0$ on $\sigma_{n,k}$, so $\Sigma_{i=1}^n Q^{ik} = P_{n-1}^{(k)}$ there. Thus, since

$$\int_{\sigma_{n,k}} P_{n-1}^{(k)} \begin{cases} dx_2 \cdots \widehat{dx_k} \cdots dx_n \\ dx_3 \cdot \cdot \cdot dx_n \end{cases} = \sum_{i \neq k} f(v_i) ,$$

we have

$$\int_{\sigma_n} (P_n f) d\mu = \sum_{i=1}^{n} f(v_i) ,$$

which completes the proof.

*Remark 4.11:* Professor E. Nelson has pointed out that a differential polynomial $P_n$ satisfying (4.21) may be obtained explicitly from the results of Section 7 of his paper [25]. Thus, letting $\nu_k$ be the unit mass on the vertex $v_k$ of $\sigma_n$, one uses Nelson's formulas (17), (24), (25) and (26) to express the Fourier transform of $\nu_k$ in terms of the Fourier transform of $\mu$. This procedure gives

$$P_n = \sum_{k=0}^{n-1} \sum_{j=0}^{n-k-1} \sum_{m=0}^{j} \binom{j}{m} \frac{(n-1)!}{(n-1-j+m)!} (-)^{n+j+k+1} \Psi^m \Phi_k \phi_{n-k-j-1} ,$$

where

$$\Psi = \sum_{i=1}^{n} x_i \frac{\partial}{\partial x_i} ,$$

$$\Phi_k = \sum_{i=1}^{n} \left(\frac{\partial}{\partial x_i}\right)^k ,$$

$$\phi_k = \sum_{i_1 < i_2 < \cdots < i_k} \frac{\partial}{\partial x_{i_1}} \cdots \frac{\partial}{\partial x_{i_k}} .$$

This polynomial does not coincide with that constructed in Lemma 4.10.

As an obvious application of Lemma 4.10:

*Corollary 4.12:* Suppose $\underline{Q}_1, \ldots, \underline{Q}_P \in F^+ \cap O$, and take any $\underline{e}_1, \ldots, \underline{e}_P$. For $c_1, \ldots, \underline{c}_P$, define $\underline{Q}(\underline{c})$, $\underline{e}(\underline{c})$ by

$$Q^{(\ell)}(\underline{c}) = \sum_{p=1}^{P} c_p Q_p^{(\ell)} ,$$

$$e^{(\ell)}(\underline{c}) = \sum_{p=1}^{P} c_p Q_p^{(\ell)} .$$

Then for $\operatorname{Re}\lambda_\ell > 2 + r_\ell + (P-1)$, we have

$$\sum_{p=1}^{P} A(\underline{\lambda}, \underline{Q}_p, \underline{e}_p) = \int_{\sigma_P} d\mu(\underline{c})(P_P A)[\underline{\lambda}, \underline{Q}(\underline{c}), e(\underline{c})] , \tag{4.23}$$

where $A$ is given by (4.19).

*Proof:* The differentiability necessary to apply Lemma 4.10 to $A$ is given in Theorem 4.5 and remarks 4.8(b), 4.9(a).

Now (4.23) gives us a method of summing the amplitudes for graphs in the theory discussed in Section 3: according to (4.20), we simply take

$$\begin{aligned} \underline{Q}_p &= \underline{Q}(G_p) , \\ \underline{e}_p &= \underline{e}(G_p) , \end{aligned} \tag{4.24}$$

in order to calculate

$$\sum_{1}^{P} A(G_p) . \tag{4.25}$$

Of course, the same would hold in any theory once the interpolating amplitude $A(\underline{\lambda}, \underline{Q}, \underline{e})$ was defined. The difficulty is that (4.23) holds only for $\operatorname{Re}\lambda_\ell > 1 + r_\ell + P$, whereas the physical point $\underline{\lambda}^0$ used in (4.20) corresponds to $\lambda_\ell = 1$. Thus we need a method of analytically continuing the integral in (4.23) to a neighborhood of $\underline{\lambda}^0$.

Let us, then, write

$$I(\underline{\lambda}) = \int_{\sigma_P} d\mu(\underline{c})(P_P A)[\underline{\lambda}, Q(\underline{c}), e(\underline{c})],$$

where Q(c), e(c) are defined using (4.24).

By Theorem 4.1 or Theorem 3.4 we know that $I(\underline{\lambda})$ has in fact a meromorphic extension to all $\underline{\lambda} \in \mathbb{C}$. Moreover, by Theorem 3.21, the expression $\mathfrak{W} I$, for any evaluator $\mathfrak{W}$, is a correctly renormalized version of (4.25). Thus the GFA has furnished a relatively simple expression for the renormalized sum of amplitudes.

Or rather, almost furnished it. We still lack an explicit formula for the integral $I(\underline{\lambda})$ which is valid near $\underline{\lambda}^0$. Now $I(\underline{\lambda})$ is really a double integral over the variables $c_1, \ldots, c_P$ and $a_1, \ldots, a_L$. It seems likely that a continuation may be constructed by some variation of the scaling methods used previously (see Theorem 4.1), but scaling both sets of variables. However, we have not found such a method to date.

# CONCLUSION

Two basic ideas have been developed in this thesis. The first is the remormalization method discussed in Chapter III, using the analytic properties of the generalized Feynman amplitude in the complex variable $\underline{\lambda}$ ; this work is completely independent of the second idea. The second idea is to use the $\underline{Q}$ and $\underline{e}$ dependence of the GFA to convert a sum of Feynman amplitudes into an integral.

The renormalization technique seems to be quite successful. Before it can be used for practical calculations, however, a way must be found to include finite renormalizations. It is possible that these may be inserted using the freedom in the choice of the generalized evaluator or in the definition of the GFA [see Remark 2.5(d)]. On the other hand, any results about the convergence or divergence of the series, even without finite renormalization, would be interesting. Analytic renormalization may also prove natural when studying zero mass particles; Shensa [30] has shown that the analytically renormalized amplitude for the "bubble" (the graph with two scalar lines joining two vertices) has a finite limit as the masses of the particles approach zero, in contrast to other methods of renormalization. Breitenlohnen and Mitter [40] have derived gauge invariant amplitudes by analytic renormalization.

Very little can be said about the value of the second idea until the integral representation has been analytically continued to the point of physical interest. We indicated in Chapter IV, Section 4, one possible approach to finding this continuation. Once it has been found, it may or may not permit estimates of size which are significantly better than those obtainable from the individual terms; if it does, some conclusions about convergence or divergence of the perturbation series could emerge.

## APPENDIX A

### *Graphs*

In this appendix we review some of the basic concepts of graph theory needed in the thesis. We then prove some familiar results on the matrices associated with Feynman graphs. The latter discussion is taken from Nakanishi, *Graph Theory and Feynman Integrals* [24].

*Definition A.1:* A *graph* G is a collection of vertices $\{V_1, \ldots, V_n\}$ and lines $\mathfrak{L}$ such that, for each line $\ell \in \mathfrak{L}$, there is assigned an initial vertex $V_{i_\ell}$ and a final vertex $V_{f_\ell}$. We always assume $V_{i_\ell} \neq V_{f_\ell}$. The term graph may also refer to the usual realization of this structure as a topological space; thus we may refer to the number of components C of a graph. We write $L = \#(\mathfrak{L})$, and let N denote the first Betti number of the graph (regarded as a simplicial complex) [19]; in the usual terminology, N is the "number of independent loops" of the graph. By Euler'a formula, $N = L - n + C$. (When several different graphs are under consideration, we write these quantities $\mathfrak{L}(G)$, C(G), etc.)

*Definition A.2:* Let G be a graph as above. A *subgraph* H of G is a subset $\{V_1', \ldots, V_m'\} \subset \{V_1, \ldots, V_n\}$ and a subset $\mathfrak{L}' \subset \mathfrak{L}$ such that $\ell \in \mathfrak{L}'$ implies $V_{i_\ell}, V_{f_\ell} \in \{V_1', \ldots, V_m'\}$. A non-empty subset $\{V_1', \ldots, V_m'\} \subset \{V_1, \ldots, V_n\}$ is called a *generalized vertex* of G. This term is also applied to the subgraph $G(V_1', \ldots, V_m')$ which consists of $\{V_1', \ldots, V_m'\}$ and *all* lines $\ell \in \mathfrak{L}$ such that $V_{i_\ell}$ and $V_{f_\ell}$ lie in $\{V_1', \ldots, V_m'\}$.

*Definition A.3:* A graph G is *one particle irreducible* (IPI) if it is connected and any subgraph obtained from G by the removal of one line is connected; otherwise, G is called *one particle reducible* (IPR). G is called *non-separable* if it is connected and if, for any vertex V of G, the topological space $G - \{V\}$ is connected. G is called *irreducible* if it is both IPI and non-separable.

*Remark A.4:* (a). The concept of irreducibility is important because a graph can be IPI and yet not non-separable, and vice versa. For example, the graph of Figure A.1 is IPI but is separable; the graph of Figure A.2 is non-separable but IPR.

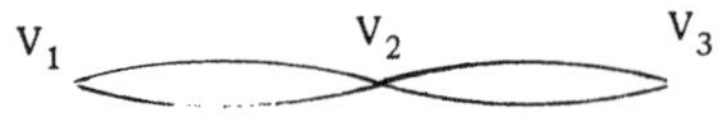

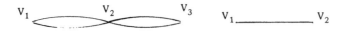

Figure A.1

Figure A.2

(b). It is easy to see that any graph G has a unique decomposition into a family of maximal irreducible subgraphs $\{G_i\}$ and single lines; in this case, $N(G) = \Sigma N(G_i)$. There is a similar decomposition of G in which the $G_i$ are maximal IPI subgraphs.

*Definition A.5:* A *tree* T is a subset of $\mathcal{L}$ such that, if $G'$ is the subgraph of G consisting of T and all vertices of lines in T, $G'$ is connected, each vertex of G is in $G'$, and $N(G') = 0$. Every tree in a connected graph G contains $n(G)-1$ lines; a disconnected graph contains no trees. For any k with $2 \leq k \leq n(G)-1$, a k-tree $T_k$ is a tree minus $k-1$ lines.

*Definition A.6:* The *incidence matrix* e of a graph G is the n by L matrix given by

$$e_i^{(\ell)} = \begin{cases} 1\ , & \text{if } V_i = V_{f_\ell}\ ; \\ -1\ , & \text{if } V_i = V_{i_\ell}\ ; \\ 0\ , & \text{otherwise.} \end{cases}$$

If T is any set of $n-1$ lines of G, and i is an integer satisfying $1 \leq i \leq n$, then $e(i,T)$ denotes the $(n-1)$-square matrix

$$e(i,T) = \{e_r^{(\ell)} \mid \ell \in T,\ i \neq r\}\ .$$

Having given the basic definitions, we now turn to the formulas for some matrices associated with a graph.

*Lemma A.7:* (a). If T is a tree, $\det[e(i,T)] = \pm 1$.
(b). If T is not a tree, $\det[e(i,T)] = 0$.

*Proof:* (a) The proof is by induction on n; the result is trivial for $n = 2$. We assume it for all $n < n_0$, and prove it for a graph with $n_0$ vertices. If only one line $\ell_0$ of T passes through $V_i$, the result follows by expanding $e(i,T)$ along the $\ell_0^{\text{th}}$ column and using the induction assumption. When several lines of T pass through $V_i$, removal of $V_i$ decomposes T into several components, to which the induction assumption may be applied. Then $e(i,T)$ is the direct sum of the matrices corresponding to each of these components.
(b). Since the matrix e satisfies $\Sigma_{i=1}^n e_i^{(\ell)} = 0$, and since every connected graph contains at least one tree, (a) implies that the incidence matrix of a connected graph has rank $n-1$. The incidence matrix for a disconnected graph is the direct sum of the matrices for its components, so that, in general, $e(G)$ has rank $n(G)-C(G)$. Now suppose T is a set of $n-1$ lines which is not a tree. Then the graph obtained by deleting all lines of G not in T is

disconnected, and hence the incidence matrix for this graph has rank at most $n-2$. This implies $\det e(i,T) = 0$.

*Definition A.8:* Suppose that G is a graph and that, to each $\ell \in \mathcal{L}(G)$, we have assigned an indeterminant $x_\ell$ ($x_\ell$ is called an *inverse Feynman parameter*). Then the matrix A (= A(G)) is defined by

$$A_{rs} = \sum_{\ell \in \mathcal{L}} e_r^{(\ell)} x_\ell e_s^{(\ell)} \,, \quad (r, s = 1, \ldots, n) \,. \tag{A.1}$$

*Lemma A.9:* For any graph G, and any i such that $1 \le i \le n$,

$$A\binom{i}{i} = \sum_T \prod_{\ell \in T} x_\ell \,, \tag{A.2}$$

the sum taken over all trees T of G. (The right hand side of (A.2) is called the *tree product sum* in G.)

*Proof:* We use the Binet-Cauchy formula: if B and C are $m \times L$ and $L \times m$ matrices, respectively, then

$$\det(BC) = \begin{cases} 0 \text{ if } L < m\,; \\ \displaystyle\sum_{k_1,\ldots,k_m} \det\{B_{ij} \mid i=1,\ldots,m,\ j = k_1,k_2,\ldots,k_m\} \\ \qquad \times \det\{C_{ji} \mid i=1,\ldots,m,\ j = k_1,k_2,\ldots,k_m\}, \text{ if } m \le L. \end{cases}$$

Here the sum is over all ordered m-tuples $1 \le k_1 < \cdots < k_m \le L$. Apply this to $B = \{e_r^{(\ell)} x_\ell \mid \ell \in \mathcal{L},\ r = 1,2,\ldots,\hat{i},\ldots,n\}$ and $C = \{e_r^{(\ell)} \mid \ell \in \mathcal{L},\ r = 1,2,\ldots,\hat{i},\ldots,n\}$, so that $m = n-1$. Then, using Lemma A.7,

$$A\binom{i}{i} = \det(BC) = \begin{cases} 0, \text{ if } L < n-1, \\ \displaystyle\sum_{T'} \Big\{[\det e(i,T)]^2 \prod_{\ell \in T'} x_\ell \Big\} = \sum_T \prod_{\ell \in T} x_\ell, \text{ if } L \ge n-1, \end{cases}$$

where $T'$ is any set of $n-1$ lines. This proves the result when $L \ge n-1$; however, when $L < n-1$, G is not connected and the right hand side of (A.2) is also zero.

*Corollary A.10:* Let $\{a_\ell \mid \ell \in \mathcal{L}\}$ be a set of strictly positive real numbers. We can evaluate A(G) at $x_\ell = a_\ell$. Then

$$A\binom{i}{i}(a_1, \ldots, a_L) = 0$$

if, and only if, G is disconnected.

*Proof:* This follows immediately from Lemma A.9.

Now that we have established (A.2), we wish to find similar expressions for $A\binom{i\,j}{i\,j}$ and $A\binom{i\,j}{i\,k}$.

*Lemma A.11:* For any graph G, and any i, j with $1 \leq i, j \leq n$,

$$A\binom{i\,j}{i\,j} = {\sum_{T_2}}' \prod_{\ell \in T_2} x_\ell \quad . \tag{A.3}$$

Here ${\sum}'_{T_2}$ is over all 2-trees $T_2$ such that $V_i$ and $V_j$ lie in different components of $T_2$.

*Proof:* Let G′ be the graph obtained from G by first deleting any lines joining $V_i$ and $V_j$, then identifying these two vertices. The matrix A(G′) is obtained from A(G) by adding the $i^{th}$ and $j^{th}$ rows and the $i^{th}$ and $j^{th}$ columns of A(G); thus $A\binom{i\,j}{i\,j}$ is given by the tree-product sum in G′. But a tree in G′ corresponds to a 2-tree in G with $V_i$ and $V_j$ in different components; this proves (A.3).

*Definition A.12:* Let G be a graph, with $\{i_1, \ldots, i_p\}$ and $\{j_1, \ldots, j_q\}$ subsets of $\{1, \ldots, n(G)\}$. Define

$$A(i_1, \ldots, i_p \,|\, j_1, \ldots, j_q) = {\sum_{T_2}}'' \prod_{\ell \in T_2} x_\ell \quad , \tag{A.4}$$

where $\Sigma''$ is over all 2-trees $T_2$ such that $\{V_{i_1}, \ldots, V_{i_p}\}$ and $\{V_{j_1}, \ldots, V_{j_q}\}$ lie on different components of $T_2$. (Note (A.4) is zero if $\{i_1, \ldots, i_p\} \cap \{j_1, \ldots, j_q\} \neq \emptyset$.) Thus (A.2) may be written

$$A\binom{i\,j}{i\,j} = A(i\,|\,j) \quad . \tag{A.5}$$

It is easily seen that, for any $1 \leq k \leq n$,

$$\begin{aligned} A(i_1, \ldots, i_p \,|\, j_1, \ldots, j_q) &= A(i_1, \ldots, i_p, k \,|\, j_1, \ldots, j_q) \\ &\quad + A(i_1, \ldots, i_p \,|\, j_1, \ldots, j_q, k) \quad . \end{aligned} \tag{A.6}$$

*Lemma A.13:* For any graph G, and any $1 \leq i, j, k \leq n$, with $i \neq j \neq k$,

$$A\binom{i\,k}{i\,j} = A(i\,|\,kj) \quad . \tag{A.7}$$

*Proof:* We first show that

$$A\binom{i\,j}{i\,j} = A\binom{i\,k}{i\,j} + A\binom{j\,k}{j\,i} \quad . \tag{A.8}$$

For simplicity, take $i = 1$, $j = 2$, $k = 3$. Then

$$A\binom{1\,2}{1\,2} = \det\{A_{rs} \,|\, r \neq 1,2;\; s \neq 1,2\} \;,$$

$$A\binom{1\,3}{1\,2} = -\det\{A_{rs} \,|\, r \neq 1,3;\; s \neq 1,2\} \;,$$

$$A\binom{2\,3}{2\,1} = -\det\{A_{rs} \,|\, r \neq 2,3;\; s \neq 1,2\} \;.$$

Thus $A\binom{1\,2}{1\,2} - A\binom{1\,3}{1\,2} - A\binom{2\,3}{2\,1}$ is the determinant of the $(n-2)$-square matrix $B$ given by

$$B_{3s} = A_{1s} + A_{2s} + A_{3s}, \qquad (s = 3, \ldots, n) ;$$

$$B_{rs} = A_{rs}, \qquad (s = 3, \ldots, n, \ r = 4, \ldots, n) .$$

Since $\Sigma_{r=1}^{n} A_{rs} = 0$, $B$ satisfies $\Sigma_{r=3}^{n} B_{rs} = 0$ and hence $\det B = 0$, proving (A.8).

Now the cyclic permutation of $i$, $j$, and $k$ in (A.8), and use of the symmetry of $A$, gives

$$A\binom{i\,k}{i\,j} = \frac{1}{2}\left[A\binom{i\,j}{i\,j} + A\binom{i\,k}{i\,k} - A\binom{j\,k}{j\,k}\right] .$$

Similar manipulations of (A.6) yield

$$A(i\,|\,k\,j) = \frac{1}{2}[A(i\,|\,j) + A(i\,|\,k) - A(j\,|\,k)] ,$$

and these two equations plus (A.5) prove (A.7).

We generalize Definition A.12 to

*Definition A.14:* Let $G$ be a graph, and let $U_1, \ldots, U_k$, $W_1, \ldots, W_k$ be non-empty subsets of $\{1, \ldots, n(G)\}$. Let $G'$ be a graph obtained by adding certain lines to $G$, specifically, $\mathcal{L}(G') = \mathcal{L}(G) \cup \bigcup_{i=1}^{k} \mathcal{L}_i$, where, for every pair $(r, s) \subset U_i \times W_i$, $\mathcal{L}_i$ contains a line $\ell$ with $V_{i_\ell} = V_r$, $V_{f_\ell} = V_s$. Then define

$$A(U_1\,|\,W_1, U_2\,|\,W_2, \ldots, U_k\,|\,W_k) = \sum_{T_{k+1}} \prod_{\ell \in T_{k+1}} x_\ell ,$$

where the sum is over all $(k+1)$-trees $T_{k+1}$ in $G$ such that $T_{k+1} \cup \{\ell_1, \ldots, \ell_k\}$ is a tree in $G'$, for any $\ell_i \in \mathcal{L}_i$.

Note that $A(U_1\,|\,W_1, \ldots, U_k\,|\,W_k)$ is zero if $U_i \cap W_i \neq \emptyset$, for any $i$, or if there exist $r, s \in \{1, \ldots, n(G)\}$ with $r \in U_i \cap U_j$, $s \in W_i \cap W_j$, for some $i \neq j$.

# APPENDIX B

## *Distributions*

The first part of this appendix is devoted to a brief review of some of the basic concepts of distribution theory. Good references for this material are [12], [31], and [34]; we will follow the notation of [34]. We then give a definition of a particular family of distributions which is of interest to us, and prove some simple results about them.

*Definition B.1:* Let $R^n$ denote n-dimensional Euclidean space; for $x \in R^n$, write $\|x\| = (x_1^2 + x_2^2 + \cdots + x_n^2)^{1/2}$. Let $\mathcal{S}$ $(= \mathcal{S}(R^n))$ consist of all complex-valued $C^\infty$ functions on $R^n$, all of whose derivatives decrease at infinity faster than any power of $\|x\|$. $\mathcal{S}$ is topologized by the family of norms

$$\|f\|_m = \sup_{\Sigma \alpha_i \leq m} \sup_{x \in R^n} (1 + \|x\|^2)^m \left| \frac{\partial^{\alpha_1}}{\partial x_1^{\alpha_1}} \cdots \frac{\partial^{\alpha_n}}{\partial x_n^{\alpha_n}} f(x) \right|$$

A *tempered distribution* is a continuous linear functional on $\mathcal{S}$. The space $\mathcal{S}' = \mathcal{S}'(R^n)$ of all tempered distributions is given the weak topology: a neighborhood of zero in $\mathcal{S}'$ has the form

$$\{T \in \mathcal{S}' \mid |(T(f_i)| < \epsilon, \ i = 1, \dots, n\} ,$$

where $f_1, \dots, f_m \in \mathcal{S}$ and $\epsilon > 0$. We will consider only tempered distributions, and will usually refer simply to "distributions." Note that any locally integrable and polynomially bounded complex-valued function $g$ on $R^n$ becomes a distribution under the rule

$$g(f) = \int_{R^n} g(x) f(x) dx \quad . \tag{B.1}$$

A distribution $T \in \mathcal{S}'$ may be differentiated according to

$$\left( \frac{\partial T}{\partial x_i} \right)(f) = - T\left( \frac{\partial f}{\partial x_i} \right) \quad .$$

For any $T \in \mathcal{S}'$ there is an integer $m$ and a positive constant $C$ such that, for all $f \in \mathcal{S}$, $|T(f)| < C\|f\|_m$. A set of distributions $X$ is called *uniformly bounded* if, for some such $C$ and $m$, $|T(f)| < C\|f\|_m$ for all $f \in \mathcal{S}$ and all $T \in X$.

*Definition B.2:* Let $G$ be a symmetric real $n \times n$ matrix with $\det G = \pm 1$. We define the *Fourier transform* $\mathcal{F} : \mathcal{S} \to \mathcal{S}$ by

$$\tilde{f}(p) = (\mathcal{F}f)(p) = (2\pi)^{-n/2} \int_{R^n} f(x) e^{-ip \circ G \circ x} dx ,$$

where $p \circ G \circ x = \Sigma_{i,j=1}^{n} p_i G_{ij} x_j$. $\mathcal{F}$ is an isomorphism of $\mathcal{S}$ onto itself; its inverse is given by

$$(\mathcal{F}^{-1} f)(x) = (2\pi)^{-n/2} \int_{R^n} f(p) e^{ip \circ G \circ x} dx$$

The Fourier transform is extended to an isomorphism $\mathcal{F}: \mathcal{S}' \to \mathcal{S}'$ by the formula

$$(\mathcal{F}T)(f) = T(\mathcal{F}f) .$$

*Remark B.3:* In the thesis proper, n is always divisible by 4 ($n = 4m$), and $R^n$ is viewed as a space of m-tuples of 4-vectors. In this case we always use $G_{i\mu,j\nu} = \delta_{ij} g_{\mu\nu}$ ($i, j = 1, \ldots, m$; $\mu, \nu = 0, 1, 2, 3$).

*Definition B.4:* Let E and F be linear subspaces of $R^n$, with $R^n = E \oplus F$. For any $f \in \mathcal{S}(R^n)$ and $y \in E$, we define $f_y \in \mathcal{S}(F)$ by $f_y(z) = f(y+z)$. Then if $T \in \mathcal{S}'(E)$, $T' \in \mathcal{S}'(F)$, we define the *tensor product* $T \otimes T' \in \mathcal{S}'(R^n)$ by

$$(T \otimes T')(f) = T[T'(f_y)] .$$

This is justified because, as may be easily shown, $T'(f_y) \in \mathcal{S}(E)$. This definition extends directly to the tensor product of more than two distributions.

*Remark B.5:* (a). Suppose that the subspaces E and F of Definition 1.4 are orthogonal; suppose, moreover, that the projections $\pi_E$ and $\pi_F$ onto them commute with the quadratic form G. Then, if T and T′ are as above, we have the simple formula

$$\mathcal{F}[T \otimes T'] = (\mathcal{F}T) \otimes (\mathcal{F}T') . \tag{B.2}$$

(b). The following situation will often arise for us. Suppose that $T \in \mathcal{S}(R^n)$ is a distribution which is invariant under translation by $z \in F$; symbolically, $T(x) = T(x+z)$ for any $z \in F$, or, more formally,

$$\Sigma\, z_i \frac{\partial T}{\partial x_i} = 0$$

for any $z \in F$. Then T is really the tensor product of a distribution $T_0$ on E and the distribution 1 on F. Since the Fourier transform of a constant is a $\delta$-function, (B.2) gives

$$\mathcal{F}T = (2\pi)^{k/2} \delta_E \mathcal{F}(T_0) . \tag{B.3}$$

where k is the dimension of F. The formula (B.3) means "$(\mathcal{F}T)(f)$ is given by restricting f to E and applying $(2\pi)^{k/2}\mathcal{F}(T_0)$ to this restriction." Note that if $\{u^{(i)}\}_{i=1}^{k}$ is an orthonormal set in F, we may write $\delta_E$ in terms of the usual Dirac $\delta$-function as

$$\delta_E(x) = \prod_{i=1}^{k} \delta(x \circ u^{(i)}) \quad .$$

*Definition B.6:* Let U be a topological space, with $u \in U$. A *distribution* T(u) *depending parametrically* on u is a map $T: U \to \mathcal{S}'(R^n)$. T(u) depends *continuously* on u if this map is continuous; this is equivalent to the continuity of the numerical function $[T(u)](f)$ for each $f \in \mathcal{S}(R^n)$. Suppose U is an open subset of $C^k$; then T(u) depends *analytically* on u if $[T(u)](f)$ is analytic on U for every $f \in \mathcal{S}(R^n)$.

We now procede to the study of a particular class of distributions; these are also discussed in [12]. Let c and $\lambda$ be complex numbers, and let Q be a complex symmetric (not Hermitian) $n \times n$ quadratic form; we write $Q = Q_1 + iQ_2$, where $Q_1$ and $Q_2$ are real; similarly $c = c_1 + ic_2$. These quantities vary in a set U(a) defined by

$$U_1(a) = \{(Q_1, c_1, \lambda) | \; c_1 \leq a < 0, \; Q_1, \lambda \text{ arbitrary}\},$$

$$U_2 = \{(Q_2, c_2) | \; Q_2 > 0, \; c_2 > 0\},$$

$$U(a) = U_1(a) \times U_2 \quad .$$

Here $Q_2 > 0$ means that $Q_2$ is positive definite (similarly $Q_2 \geq 0$ means positive semi-definite). The constant a itself is not important; what is relevant is that it be strictly negative.

We now define a distribution depending parametrically on $(Q, c, \lambda) \in U(a)$.

*Definition B.7:* For $(Q, c, \lambda) \in U(a)$, and $p \in R^n$, consider the expression $(p \circ Q \circ p + c) = (\Sigma_{i,j=1}^n p_i Q_{ij} p_j + c)$. The imaginary part of this expression is always positive, so we may take $0 < \arg(p \circ Q \circ p + c) < \pi$ and define

$$(p \circ Q \circ p + c)^\lambda = \exp \lambda \{\ln |p \circ Q \circ p + c| + i \arg(p \circ Q \circ p + c)\} \quad . \tag{B.4}$$

The function (B.4) becomes a distribution in $\mathcal{S}'(R^n)$ according to (B.1); this distribution is written simply

$$(Q + c)^\lambda \quad .$$

It is easy to see that $(Q+c)^\lambda$ is continuous on U(a) and analytic in $\lambda$ for fixed Q and c. We extend these properties by

*Theorem B.8:* The distribution $(Q+c)^\lambda$ may be extended to a distribution $(Q + c + i0)^\lambda$ defined and continuous on $\overline{U(a)}$ (the closure of U(a)), and analytic there in $\lambda$ for fixed Q, c.

We remark that the difficulty in making this extension arises because, for $(Q, c, \lambda) \in \overline{U(a)}$, the expression $p \circ Q \circ p + c$ can vanish for certain values of p, so that, for $\operatorname{Re} \lambda \leq -1$, the function $(p \circ Q \circ p + c)^\lambda$ is not locally integrable. The proof will also be made somewhat more elaborate by the necessity to prove the continuity as well as existence of the extension. We first prove two lemmas.

*Lemma B.9:* Let $K \subset \overline{U(a)}$ be compact. Then for $(Q, c, \lambda) \in K \cap U(a)$, $(Q+c)^\lambda$ is uniformly Cauchy in the variables $(Q_2, c_2)$; specifically, there exist constants $m \geq 0$, $M > 0$ such that, for $f \in \mathcal{S}(R^n)$,

$$\text{(B.5)} \qquad |(Q_1 + iQ_2 + c_1 + ic_2)^\lambda(f) - (Q_1 + iQ_2' + c_1 + ic_2')^\lambda(f)| < M\|f\|_m(\|Q - Q'\| + |c_2 - c_2'|) .$$

*Proof:* Case 1. Suppose $(Q, c, \lambda) \in K$ implies $\mathrm{Re}\,\lambda > 1$. Then the derivative of $(p \circ Q \circ p + c)^\lambda$ with respect to $(Q_2)_{ij}$ or $c_2$ is uniformly polynomially bounded in $p$ for $(Q, c, \lambda) \in K$, so the lemma is true.

Case 2. Suppose $K$ is arbitrary. Let $\psi$ be a fixed $C^\infty$ function on $R$ with $0 \leq \psi(x) \leq 1$ and

$$\psi(x) = \begin{cases} 1, & \text{if } x > -2a/3 \\ 0, & \text{if } x < -a/3 \end{cases} .$$

Define operators $\mathcal{T}_1(Q_1)$, $\mathcal{T}_2(Q_1)$: $\mathcal{S} \to \mathcal{S}$ by

$$(\mathcal{T}_1 f)(p) = \psi(p \circ Q_1 \circ p) f(p)$$

$$(\mathcal{T}_2 f)(p) = [1 - \psi(p \circ Q_1 \circ p)] f(p) .$$

These operators are uniformly bounded on $K$, so it suffices to prove (B.5) with $f$ replaced by $\mathcal{T}_1 f$ and $\mathcal{T}_2 f$. Since $(p \circ Q \circ p + c)$ cannot vanish for $p \in \mathrm{Supp}\,\mathcal{T}_2 f$, the proof for $\mathcal{T}_2 f$ is precisely as in Case 1.

Now $(p \circ Q \circ p + c)^\lambda$ satisfies

$$\text{(B.6)} \qquad \sum_{i=1}^{n} p_i \frac{\partial}{\partial p_i} (p \circ Q \circ p + c)^{\lambda+1} = 2(\lambda+1)(p \circ Q \circ p)(p \circ Q \circ p + c)^\lambda ,$$

so that defining the operator $\mathcal{Q}(Q)$ on $\mathcal{S}$ by

$$(\mathcal{Q} f)(p) = -\sum_{i=1}^{n} \frac{\partial}{\partial p_i} \left[ \frac{p_i f(p)}{p \circ Q \circ p} \right]$$

we have from (B.6)

$$\text{(B.7)} \qquad (Q + c)^\lambda(\mathcal{T}_1 f) = \frac{1}{2(\lambda+1)} (Q + c)^{\lambda+1} [\mathcal{Q}(Q) \mathcal{T}_1 f]$$

or iterating (B.7) $k$ times,

$$\text{(B.8)} \qquad (Q + c)^\lambda(\mathcal{T}_1 f) = \frac{1}{2^k(\lambda+1)\cdots(\lambda+k)} (Q + c)^{\lambda+k} [\mathcal{Q}(Q)^k \mathcal{T}_1 f] .$$

All this is permissible because, for $p \in \mathrm{supp}\,\mathcal{T}_1 f$, $\mathrm{Re}(p \circ Q \circ p) > -a/3$; thus $\mathcal{Q}(Q)$ is uniformly bounded on the functions $\mathcal{T}_1 f$. Therefore, if we take $k$ such that $(Q, c, \lambda) \in K \Longrightarrow \mathrm{Re}\,\lambda > -k+1$, (B.8) plus Case 1 prove the lemma. The only difficulty arises when $\lambda$ is a negative integer, in which case (B.8) breaks down. However, since $(Q+c)^\lambda$ is analytic in $\lambda$, we can apply the maximum modulus principle to, say, a disc of radius ½ centered at $\lambda$; since (B.5) holds on the boundary of the disc, the same is true in the interior. This completes the proof.

*Lemma B.10:* Let $X$ and $Y$ be metric spaces, with metrices $d$ and $d'$ respectively, and let $V$ be a dense subset of $Y$. If $g$ is a function continuous on $X \times V$ and satisfying, for some $M > 0$,

$$|g(x,y) - g(x,y')| < M d'(y,y') \quad , \tag{B.9}$$

then $g$ has a continuous extension to $X \times Y$.

*Proof:* By (B.9) we can define $g$ on $X \times Y$ by

$$g(x,y) = \lim_{\substack{y' \to y \\ y' \in V}} g(x,y') \ . \tag{B.10}$$

We now show that $g$ is continuous at $(x,y) \in X \times Y$. Given $\epsilon > 0$, choose $y' \in V$ with $d'(y,y') < \epsilon/4M$. Since $g(x,y')$ is continuous in $x$, there is a $\delta > 0$ such that $d(x,x') < \delta$ implies $|g(x,y') - g(x',y')| < \epsilon/4$. Hence, whenever $d(x,x') < \delta$ and $d'(y,y'') < \epsilon/4M$, we have

$$\begin{aligned}|g(x,y) - g(x',y'')| &\le |g(x,y) - g(x,y')| + |g(x,y') - g(x',y')| + |g(x',y') - g(x',y'')| \\ &\le \epsilon/4 + \epsilon/4 + \epsilon/2 = \epsilon \ ,\end{aligned}$$

which completes the proof. Note the importance of the uniformity of (B.9) in $x$.

*Proof of Theorem B.8:* Let $X_m$ and $Y_m$ be compact subsets of $U_1(a)$ and $\overline{U}_2$, respectively, such that $U_1(a) = \cup_{m=1}^{\infty} X_m$, $\overline{U}_2 = \cup_{m=1}^{\infty} Y_m$. Then the existence and continuity of $(Q+c+i0)^{\lambda}$ follow by applying Lemma B.10 successively to $X_1 \times Y_1$, $X_2 \times Y_2$, etc; the necessary estimate (B.9) is precisely (B.5). The analyticity of the limit in $\lambda$ is proved by noting that, according to (B.10), $(Q+c+i0)^{\lambda}$ is for fixed $Q$ and $c$ the uniform limit on compact sets (in $\lambda$) of analytic functions, q. e. d.

We now discuss the Fourier transform of the distribution $(Q+c)^{\lambda}$. This question is treated carefully in Gel'fand and Shilov [12], so we omit the proof of the following theorem. Equation (B.11) differs slightly from the formula of Gel'fand due to the difference in the definitions of the Fourier transform and the notation for the distributions.

Note that for $(Q,c,\lambda) \in U(a)$, the quadratic form $Q$ has an inverse; moreover, the imaginary part of $Q^{-1}$ is negative definite.

*Theorem B.11:* (a). Suppose $(Q,c,\lambda) \in U(a)$. Then

$$\mathcal{F}^{-1}(Q+c)^{\lambda} = \frac{2^{\lambda+1} e^{\lambda\pi i} b^{n/2+\lambda}}{\Gamma(-\lambda)[\det(-Q)]^{1/2}} \frac{K_{n/2+\lambda}[bP^{1/2}]}{[P^{1/2}]^{n/2+\lambda}}$$

where (i) $K$ is the "Bessel function of imaginary argument" [37];

(ii) $b = \sqrt{-c}$, defined so that $\operatorname{Re} b > 0$;

(iii) $P$ is the quadratic form $P = -G \circ Q^{-1} \circ G$, and $P^{1/2}$ denotes $[x \circ P \circ x]^{1/2}$. Since when we write $P = P_1 + iP_2$ we have $P_2$ positive definite, we take $0 < \arg[x \circ P \circ x]^{1/2} < \pi/2$, and use this to define $[P^{1/2}]^{n/2+\lambda}$ as in (B.4);

(iv) we use the simple connectivity of the region $\{Q \mid Q = Q_1 + i\,Q_2,\ Q_1 < 0$ or $Q_2 > 0\}$ to define the sign of $[\det(-Q)]^{1/2}$, taking the sign as positive for $Q = Q_1 < 0$.

(b). Now suppose $Q_1$ is invertible. Then as $Q_2 \to 0$, the quadratic form $P$ has the limit $-G \circ Q_1^{-1} \circ G$; we denote this $P + i0$. Then (B.11) becomes

$$\text{(B.12)} \qquad \mathcal{F}^{-1}(Q + c + i0)^{\lambda} = \frac{2^{\lambda+1}\, e^{\lambda\pi i}\, b^{n/2+\lambda}}{\Gamma(-\lambda)\,[\det(-Q)]^{1/2}} \; \frac{K_{n/2+\lambda}[b\,(P+i0)^{1/2}]}{[(P+i0)^{1/2}]^{n/2+\lambda}} \,.$$

In this formula $\det(-Q) \neq 0$ because $Q_1$ is invertible; $[\det(-Q)]^{1/2}$ is defined as in (iv) above, using the fact that $Q$ lies on the boundary of the region discussed there.

## APPENDIX C

### *The Free Field*

(A). PRELIMINARIES.

We will study fields $\phi(x)$ which satisfy the equation

(C.1) $$(-i\gamma^{\mu} \frac{\partial}{\partial x^{\mu}} + m)\phi(x) = 0 .$$

Here $\phi$ is a mapping $\phi$: $R^4 \to C^M$, and $\{\gamma^{\mu}\}$ $(\mu = 0, 1, 2, 3)$ are $M \times M$ matrices. Thus, we consider elements in $C^M$ as column vectors; if $u = \begin{pmatrix} u_1 \\ \vdots \\ u_m \end{pmatrix}$ is such a vector, we write $u^{\dagger} = (\bar{u}_1, \ldots, \bar{u}_M)$; the usual scalar product $(u, v) = \Sigma_{i=1}^{M} \bar{u}_i v_i$ in $C^M$ becomes $(u, v) = u^{\dagger}v$. Similarly, $A^{\dagger}$ denotes the Hermitian adjoint of the $M \times M$ matrix A. We will adopt the notation of Feynman and, for any 4-vector $a \in C^4$, write $\not{a} = a_{\mu}\gamma^{\mu}$. We also suppose that we are given an $M \times M$ matrix representation $A \to S(A)$ of $SL(2, C)$, which is either a pure spinor or pure tensor representation.

Such equations are discussed from a classical (i.e., unquantized) point of view in various works; see, for example, [6, 11]. The quantization of the equation is discussed in [35]; our treatment, however, resembles more closely the traditional methods of quantizing the Dirac equation [32]. A somewhat different point of view in the quantization of free fields is given in [38].

*Remark:* Suppose that $v = (v_0, \vec{v})$ is a real 4-vector with $v^2 = 1$. The "boost" $\Lambda(v)$ is the Lorentz transformation which gives particles at rest a velocity $\vec{v}$, specifically,

$$\Lambda(v)^i{}_j = \delta_{ij} + \frac{v_i v_j}{v_0 + 1}$$

$$\Lambda(v)^i{}_0 = \Lambda(v)^0_i = v_i$$

$$\Lambda(v)^0_0 = v_0 .$$

Then we define $A(v) \in SL(2, C)$ (also a boost) so that $\Lambda[A(v)] = \Lambda(v)$ and so that $\mathrm{Tr}\, A(v) > 0$; this defines $A(v)$ uniquely.

We make the following assumptions about the matrices $\{\gamma^{\mu}\}$ and $S(A)$:

(1).

(C.2) $$S(A)^{-1} \gamma^{\mu} S(A) = \Lambda^{\mu}{}_{\nu}(A)\gamma^{\nu} .$$

This is motivated as follows: the set of maps $\phi: R^4 \to C^M$ is a vector space. If we define a representation $S'$ of inhomogeneous $SL(2, C)$ on this space by

$$[S'(a, A)\phi](x) = S(A)\phi(A^{-1}(x-a)) , \tag{C.3}$$

then (C.2) guarantees that $S'$ preserves solutions of (C.1).

(2). There is an invertible Hermitian matrix $\eta$ such that

$$\eta\gamma^\mu\eta^{-1} = \gamma^{\mu\dagger} \qquad (\mu = 0, 1, 2, 3), \tag{C.4}$$

$$\eta S(A)\eta^{-1} = S(A)^\dagger \qquad (A \in SL(2, C)) . \tag{C.5}$$

For any $u \in C^M$ we write $\tilde{u} = u^\dagger\eta$; then (C.5) guarantees that the inner product $<u, v> = \tilde{u}v$ on $C^M$ is invariant under $S(A)$.

(3). There is a matrix $C$ satisfying

$$C^{-1}\gamma^\mu C = -\bar{\gamma}^\mu , \tag{C.6}$$

$$C^{-1}S(A)C = \overline{S(A)} , \tag{C.7}$$

$$\bar{C} = C^{-1} \tag{C.8}$$

$$C^\dagger\eta C = (-)^\sigma\bar{\eta} \tag{C.9}$$

where $\sigma$ is zero (one) when $S$ is a tensor (spinor) representation.

(4). The matrix $\gamma^0$ has only $0$ and $\pm 1$ as eigenvalues. (This guarantees that solutions of (C.1) actually represent particles of mass $m$; see [11] and below.)

(5). If $u$ is any non-zero eigenvector of $\gamma^0$ with eigenvalue $\lambda = 1$, we have $\tilde{u}u > 0$. (This corresponds to the requirement in [11] that a tensor field has positive energy and a spinor field positive charge.)

(B). SOLUTIONS OF THE FIELD EQUATIONS.

We now study the classical (i.e., unquantized) solutions of (C.1); these will be used to construct the quantized version of the theory. It is convenient to work with Fourier transforms so that, defining

$$\phi(p) = (2\pi)^{-2}\int_{R^4}\phi(x)e^{ip\cdot x}\,dx ,$$

equation (C.1) becomes

$$(\not{p} - m)\,\phi(p) = 0 . \tag{C.10}$$

We begin our study of (C.10) by deriving some properties of the $\gamma$-matrices.

By the theorem of the Jordan canonical form, $C^M$ has a basis of vectors $\{u^\lambda_{j,k} \mid \lambda = 0, \pm 1; j = 1,\ldots, j(\lambda),\ k = 0, 1, \ldots, k(j, \lambda)\}$ satisfying

$$\gamma^0 u^\lambda_{jk} = \begin{cases} \lambda u^\lambda_{j,k} & \text{if } k = 0 \qquad \text{(C.11a)} \\ \lambda u^\lambda_{j,k} + u^\lambda_{j,k-1} & \text{if } k > 0 . \qquad \text{(C.11b)} \end{cases}$$

If we take $k = 1$ and $\lambda = 1$, multiply (C.11b) by $\tilde{u}^{\lambda}_{j,0}$, and use (C.4), we find $\tilde{u}^{\lambda}_{j,0}\, u^{\lambda}_{j,0} = 0$, contradicting assumption (5). Thus $k$ cannot be 1 for $\lambda = \pm 1$, i.e., $k(j, \pm 1) = 0$ for all $j$, and hence $\gamma^0$ satisfies the equation

$$\text{(C.12)} \qquad [(\gamma^0)^2 - 1](\gamma^0)^q = 0 \ ,$$

with $q = \max_j k(j,0) + 1$.

Now, it follows from (C.2) that for any 4-vector $a$,

$$\text{(C.13)} \qquad S(A)\, \not{a}\, S(A)^{-1} = (\not{Aa}) \ .$$

Suppose $a$ is timelike, and let $b = a/\sqrt{a^2}$. Then multiplying (C.12) on the left by $S[A(b)]$ and on the right by $S[A(b)]^{-1}$, we have

$$\text{(C.14)} \qquad (\not{a}^2 - a^2)\not{a}^q = 0 \ .$$

This is the basic equation satisfied by the $\gamma$-matrices.

We now discuss a statement made earlier to the effect that solutions of (C.1) represent particles of mass $m$. Define, for any 4-vector $a$,

$$\text{(C.15)} \qquad d(\not{a}) = (\not{a} + m)\left(\frac{\not{a}}{m}\right)^q - \left[\frac{a^2 - m^2}{m}\right] \sum_{i=0}^{q-1} \left(\frac{\not{a}}{m}\right)^i \ ,$$

and note that $d(\not{a})(\not{a} - m) = a^2 - m^2$. Thus if we apply $d(i\not{\partial})$ to (C.1) we find that every solution $\phi$ of (C.1) also satisfies the Klein-Gordon equation

$$(\partial_\mu \partial^\mu + m^2)\phi(x) = 0 \ ,$$

which is what we claimed. This means that any solution $\phi(p)$ of (C.10) will vanish except for $p^2 = m^2$.

Suppose, then, that $\vec{p}$ is any 3-vector and write $p^0 = (m^2 + |\vec{p}|^2)^{1/2}$ $(p^0 > 0)$. We first look for solutions of the equation

$$\text{(C.16)} \qquad (\not{p} - m)u = 0$$

with $u \in C^M$. These are easily constructed as follows: let $\{u_i\}_{i=1}^m$ be a complete set of eigenvectors of $\gamma^0$ with eigenvalue 1, normalized to $\tilde{u}_i u_j = \delta_{ij}$ [this is possible by assumption (5)]. Then define

$$u_i(p) = S[A(\tfrac{p}{m})]\, u_i(0) \ ;$$

it follows from (C.13) that $u_i(p)$ is a solution of (C.16). Since for any solution $u$ of (C.15), $S[A(\frac{p}{m})]^{-1}u$ is an eigenvector of $\gamma^0$ with eigenvalue one, the set $\{u_i(p)\}_{i=1}^m$ forms a complete set of solutions of (C.16). Notice that $\tilde{u}_i(p)u_j(p) = \delta_{ij}$.

Now define $v_i(\vec{p}) = C\, u_i(\vec{p})$. From (C.6) and (C.16) we have

$$\text{(C.17)} \qquad (\not{p} + m)v_i(\vec{p}) = 0 \ ,$$

where again $p = (p^0, \vec{p})$. On the other hand, we are looking for solutions of (C.10); thus, if we let $p' = (-p^0, \vec{p})$, we have from (C.17)

$$(\not{p}' - m) v_i(-\vec{p}) = 0 \ .$$

Thus the set $\{v_i(-\vec{p})\}$ is, for fixed $\vec{p}$, a complete set of negative energy solutions of (C.10). Notice that, from (C.9), $\tilde{v}_i(p) v_j(p) = (-)^\sigma \delta_{ij}$. We also have, from (C.7),

$$v_i(p) = S[A(\tfrac{p}{m})] v_i(0) \ . \tag{C.18}$$

We can now write down the general solution of (C.1):

$$\phi(x) = (2\pi)^{-2} \int d^4k\, e^{-ik\cdot x} \phi(k) = (2\pi)^{-2} \int d^4k\, e^{-ik\cdot x} \delta(k^2 - m^2) \hat{\phi}(k) \ ,$$

where $\hat{\phi}$ is defined on the hyperboloids $k^2 = m^2$. Writing $\hat{\phi}(k)|_{k_0 > 0} = \phi^{(1)}(\vec{k})$; $\phi(k)|_{k_0 < 0} = \phi^{(2)}(-\vec{k})$, we have

$$\phi(x) = \frac{1}{2(2\pi)^2} \int d\Omega(\vec{p}) \{\phi^{(1)}(\vec{p}) e^{-ip\cdot x} + \phi^{(2)}(\vec{p}) e^{ip\cdot x}\} \ , \tag{C.19}$$

where in (C.19) $p^0 = [m^2 + |\vec{p}|^2]^{1/2}$, as before. Finally, we require $\phi^{(1)}$ and $\phi^{(2)}$ to be of the form

$$\phi^{(1)}(\vec{k}) = \sum_{i=1}^{m} f_i(\vec{k}) u_i(\vec{k})$$

$$\phi^{(2)}(\vec{k}) = \sum_{i=1}^{m} g_i(\vec{k}) v_i(\vec{k}) \ .$$

Before turning to the quantized theory we make a final observation. If $u, v \in C^M$, then $u \otimes \tilde{v}$ is the operator on $C^M$ given by $(u \otimes \tilde{v})(w) = (\tilde{v} w) u$. Now (in the notation of page 105) we have

$$\tilde{u}^{\lambda}_{j,k}\, u^{\lambda'}_{j,k} = 0 \ ,$$

for $\lambda \neq \lambda'$. Using this it is easy to verify the equation

$$\sum_{i=1}^{m} u_i(0) \otimes \tilde{u}_i(0) = \tfrac{1}{2} (\gamma^0 + 1)(\gamma^0)^q \tag{C.20}$$

by applying both sides to all vectors in the basis $\{u^{\lambda}_{jk}\}$. Multiplying (C.20) on the left by $S[A(\frac{p}{m})]$ and on the right by $S[A(\frac{p}{m})]^{-1}$ gives

$$\sum_{i=1}^{m} u_i(\vec{p}) \otimes \tilde{u}_i(\vec{p}) = \frac{1}{2m} (\not{p} + m)(\tfrac{\not{p}}{m})^q \ , \tag{C.21a}$$

$$= \frac{1}{2m} d(\not{p}) \ , \tag{C.21b}$$

where we have used $p^2 = m^2$ in obtaining (C.21b). Similarly, one proves

$$\sum_{i=1}^{m} v_i(\vec{p}) \otimes \tilde{v}_i(\vec{p}) = \frac{(-)^\sigma}{2m} d(-\not{p}) \quad . \tag{C.22}$$

(C). QUANTIZING THE THEORY

(1). Fock space.

We begin by constructing the Fock space based on the positive energy solutions of (C.1). Thus let $\mathcal{H}_0 = \mathbb{C}$, and for any $k > 0$, define $\mathcal{H}_k$ to be all functions $\Phi^{(k)}: \mathbb{R}^{3k} \to \mathbb{C}^{Mk}$ such that

(a) $$\Phi^{(k)} (\vec{p}_1 \cdots \vec{p}_k) = \sum_{i_1 \cdots i_k = 1}^{m} \Phi^{(k)}_{i_1 \cdots i_k}(\vec{p}_1 \cdots \vec{p}_k) u_{i_1}(\vec{p}_1) \otimes \cdots \otimes u_{i_k}(\vec{p}_k) \ ;$$

(b) $\Phi^{(k)}_{i_1 \cdots i_k}(\vec{p}_1 \cdots \vec{p}_k)$ is totally symmetric or antisymmetric under permutation of the i's and p's, for S a tensor or spinor representation, respectively;

(c). Defining $\tilde{\Phi}^{(k)} = \Phi^{(k)\dagger}(\eta \otimes \cdots \otimes \eta)$, we have

$$\int_{R^{3n}} d\Omega(\vec{p}_1) \cdots d\Omega(\vec{p}_k) \tilde{\Phi}^{(k)}(\vec{p}_1 \cdots \vec{p}_k) \Phi^{(k)}(\vec{p}_1 \cdots \vec{p}_k) < \infty \ .$$

$\mathcal{H}_k$ is equipped with the scalar product

$$(\Psi^{(k)}, \Phi^{(k)}) = \int_{R^{3n}} d\Omega(\vec{p}_1) \cdots d\Omega(\vec{p}_k) \tilde{\Psi}^{(k)}(\vec{p}_1 \cdots \vec{p}_k) \Phi^{(k)}(\vec{p}_1 \cdots \vec{p}_k)$$

and a representation $U_k(a, A)$ of inhomogeneous $SL(2, \mathbb{C})$ given by

$$U_k(a, A)\Phi^{(k)}(p_1 \cdots p_k) = e^{i a \cdot \Sigma_1^k p_i} S(A) \otimes \cdots \otimes S(A) \Phi^{(k)}(A^{-1}p_1 \cdots A^{-1}p_k) \tag{C.23}$$

which is easily seen to be unitary. (In (C.23) we have written $\Phi^{(k)}$ as if it depended on the 4-momentum $[(|\vec{p}|^2 + m^2)^{1/2}, \vec{p}]$ for notational convenience.) Finally, we let $\mathcal{H} = \mathcal{H}_0 \oplus \mathcal{H}_1 \oplus \cdots$; $\mathcal{H}$ is equipped with the representation $U(a, A) = 1 \oplus U_1(a, A) \oplus \cdots$ and a vector $\Phi \in \mathcal{H}$ is written $\Phi = (\Phi^{(0)}, \Phi^{(1)} \cdots)$.

We now define the creation and annihilation operators $a_i^*(\vec{p})$, $a_i(\vec{p})$ $(i = 1 \cdots m)$ by the formulas

$$[a_i(\vec{p})\Phi]^{(k)}_{i_1 \cdots i_k}(\vec{p}_1 \cdots \vec{p}_k) = \sqrt{k+1}\, \Phi^{(k+1)}_{i, i_1 \cdots i_k}(\vec{p}, \vec{p}_1 \cdots \vec{p}_k) \ , \tag{C.24}$$

$$[a_i^*(\vec{p})\Phi]^{(k)}_{i_1 \cdots i_k}(\vec{p}_1 \cdots \vec{p}_k) = \frac{1}{\sqrt{k}} \sum_{j=1}^{k} (-)^{\sigma(j+1)} \delta_{ii_j} p^0 \delta(\vec{p} - \vec{p}_j) \times \Phi^{(k-1)}_{i_1 \cdots \hat{i}_j \cdots i_k}(\vec{p}_1 \cdots \hat{\vec{p}}_j \cdots \vec{p}_k) \ . \tag{C.25}$$

(Recall that $\sigma$ is zero (one) for S a tensor (spinor) representation.) We will sometimes write $a(\vec{p})$ as $a(p)$, where $p = (p^0, \vec{p})$. Purists who are unhappy with (C.25) may define operators $a_i(f)$ and $a_i^*(f)$ by smearing (C.24) and (C.25) with a test function $f(\vec{p})$ in the obvious way. From (C.24) and (C.25) we have the commutation relations

$$[a_i(\vec{p}), a_j^*(\vec{q})]_{\pm} = p^0 \delta_{ij} \delta(\vec{p}-\vec{q}) \ , \tag{C.26}$$

$$[a_i(\vec{p}), a_j(\vec{q})]_{\pm}' = 0 \ , \tag{C.27}$$

where the sign on the commutator is taken to be $(-)^{\sigma+1}$.

[We digress a moment to describe all this in the usual intuitive language. Take $\omega \in \mathcal{H}_0$, with $(\omega, \omega) = 1$ as the vacuum of the theory. Then the "vectors"

$$|\vec{p}_1, i_1, \cdots, \vec{p}_k, i_k> = \frac{1}{\sqrt{k!}} a_{i_1}^*(\vec{p}_1) \cdots a_{i_k}^*(\vec{p}_k)\, \omega$$

represent states with k particles in the one-particle states $u_{i_1}(\vec{p}_1), \ldots, u_{i_k}(\vec{p}_k)$. If $\Phi \in \mathcal{H}$, its wave functions are given by

$$\Phi^{(k)}_{i_1 \cdots i_k}(\vec{p}_1 \cdots \vec{p}_k) = <\vec{p}_1, i_1 \cdots \vec{p}_k, i_k | \Phi> \ .$$

Formulae (C.26) and (C.27) then imply

$$a_i^*(\vec{p})|\vec{p}_1, i_1 \cdots \vec{p}_k, i_k> = \sqrt{k+1}\, |\vec{p}, i, \vec{p}_1, i_1, \ldots, \vec{p}_k, i_k>$$

$$a_i(\vec{p})|\vec{p}_1, i_1 \cdots \vec{p}_k, i_k> = \frac{1}{\sqrt{k}} \sum_{j=1}^{k} (-)^{\sigma(j+1)} p^0 \delta(\vec{p}-\vec{p}_j) \delta_{ii_j} \times |\vec{p}_1, i_1 \cdots \widehat{\vec{p}_j, i_j} \cdots p_k, i_k>$$

from which (C.24) and (C.25) follow.]

The transformation properties of the $a_i(\vec{p})$ may be derived from (C.23) and (C.24). In particular, one may show

$$\sum_{i=1}^{m} U(a, A) a_i(p) u_i(p) U(a, A)^{-1} = \sum_{i=1}^{m} e^{-a \cdot Ap} a_i(Ap) S(A)^{-1} u_i(Ap) \tag{C.28}$$

a relation we will need later. Taking the adjoint (as operators on $\mathcal{H}$) of (C.28) and using (C.7) gives

$$\sum_{i=1}^{m} U(a, A) a_i^*(\vec{p}) v_i(p) U(a, A)^{-1} = \sum_{i=1}^{m} e^{-a \cdot Ap} a_i^*(Ap) S(A)^{-1} v_i(Ap) \ . \tag{C.29}$$

We now discuss the two essentially different ways of quantizing (C.1).

(2). Non-self-charge-conjugate theory.

The Hilbert space for this theory is given by $\mathcal{H}^N = \mathcal{H}^{(a)} \otimes \mathcal{H}^{(b)}$, where $\mathcal{H}^{(a)}$ and $\mathcal{H}^{(b)}$ are two copies of the Fock space given above. Thus $\mathcal{H}^{(a)}$ is the space of (say) particles, $\mathcal{H}^{(b)}$ the space of antiparticles. The vacuum $\Psi_0$ of the theory is any vector of norm one lying in $\mathcal{H}_0^{(a)} \otimes \mathcal{H}_0^{(b)}$; the representation of the inhomogeneous SL(2, C) is the tensor product of the representations in $\mathcal{H}^{(a)}$ and $\mathcal{H}^{(b)}$. We denote the creation and destruction operators in $\mathcal{H}^{(a)}$ and $\mathcal{H}^{(b)}$ by $a_i^*(\vec{p})$, $a_i(\vec{p})$ and $b_i'^*(\vec{p})$, $b_i'(\vec{p})$ respectively.

These operators are not suitable for construction of fields in the case $\sigma = 1$ because $a$ and $a^*$ commute with $b'$ and $b'^*$ rather than anticommuting. We remedy this by defining new operators as follows. An element $\Phi \in \mathcal{H}^N$ has a decomposition

$$\Phi = \sum_{r,s=0}^{\infty} \Phi^{(r,s)} ,$$

where $\Phi^{(r,s)}$ is a state with $r$ particles and $s$ antiparticles. Then define operators $b_i(\vec{p})$ and $b_i^*(\vec{p})$ by

$$[b_i(\vec{p})\Phi]^{(r,s)} = (-)^{\sigma \cdot r}[b_i'(\vec{p})\Phi]^{(r,s)} ,$$

and similarly for $b_i^*(\vec{p})$. The $b$ operators have the same commutation relations and Lorentz transformation properties as the $b'$ operators, but they now have appropriate (anti) commutation relations with the $a$'s.

The field $\phi(x)$ is now defined by

$$\phi(x) = (2\pi)^{-2} \int d\Omega(\vec{p}) \sum_{i=1}^{m} \{a_i(\vec{p})\, u_i(\vec{p})\, e^{-ip \cdot x} + b_i^*(\vec{p})\, v_i(\vec{p})\, e^{ip \cdot x}\} \tag{C.30}$$

Note the similarity of this equation to (C.19). For completeness' sake we write this definition in the precise sense of [34]: for $f \in \mathcal{S}(R^4)$, define

$$\tilde{f}(p) = (2\pi)^{-2} \int_{R^4} f(x)\, e^{-ip \cdot x} ,$$

and

$$\left\{ \begin{aligned} f^{(+)}(\vec{p}) &= f(p^0, \vec{p}) \\ f^{(-)}(\vec{p}) &= f(-p^0, \vec{p}) \end{aligned} \right\} ,$$

with $p_0 > 0$, $(p^0)^2 = m^2 + |\vec{p}|^2$.

For $g \in \mathcal{S}(R^3)$, define

$$a_i(g) = \int a_i(\vec{p})\, g(\vec{p})\, d\Omega(\vec{p}) .$$

Then the precise version of (C.30) is

$$\phi_\alpha(f) = \sum_{i=1}^{m} \{a_i(f^{(+)} u_{i,\alpha}) + b_i^*(f^{(-)} v_{i,\alpha})\} ,$$

for any $1 \leq \alpha \leq M$.

The transformation properties of $\phi$ follow immediately from (C.28) and (C.29):

(C.31)
$$U(a, A)\phi(x)U(a, A)^{-1} = \phi(Ax + a) \;.$$

Moreover, it follows from (C.16) and (C.17) that $\phi(x)$ obeys (C.1). In discussing commutation relations, it is useful to consider the row vector $\phi^{\dagger}(x) = (\phi_1{}^*(x), \ldots, \phi_M^*(x))$ or more precisely the vector $\bar{\phi}(x) = \phi^{\dagger}(x)\eta$. It is clear from (C.27) that each component of $\phi$ commutes (or anticommutes, depending on $\sigma$) with itself. Then

$$[\phi(x), \bar{\phi}(y)]_{\pm} = (2\pi)^{-4} \int d\Omega(\vec{p})\, d\Omega(\vec{q}) \sum_{i,j=1}^{m} \{u_i(\vec{p}) \otimes \bar{u}_j(\vec{q})\,[a_i(\vec{p}), a_j^*(\vec{q})]_{\pm}$$
$$\times\, e^{-i[p\cdot x - q\cdot y]} + v_i(p) \otimes \bar{v}_j(q)\,[b_i^*(\vec{p}), b_i(\vec{q})]_{\pm}\, e^{i(p\cdot x - q\cdot y)}\} \;.$$

Using (C.26) and doing the $\vec{q}$-integral and j-sum gives

$$[\phi(x), \bar{\phi}(y)]_{\pm} = (2\pi)^{-4} \int d\Omega(\vec{p}) \sum_{i=1}^{m} \{[u_i(\vec{p}) \otimes \bar{u}_i(\vec{p})]\, e^{-ip\cdot(x-y)}$$
$$+ (-)^{\sigma+1}[v_i(\vec{p}) \otimes \bar{v}_i(\vec{p})]\, e^{ip\cdot(x-y)}\} \;;$$

if we now insert (C.21 b) and (C.22) and convert $d(\not{p})$ into an x-derivative, we find

(C.32)
$$[\phi(x), \bar{\phi}(y)]_{\pm} = \frac{d(i\not{\partial}_x)}{2m(2\pi)^4} \int d\Omega(\vec{p})\,[e^{-ip\cdot(x-y)} - e^{ip\cdot(x-y)}]$$
$$= \frac{i}{2\pi}\,\frac{d(i\not{\partial}_x)}{m}\,\Delta(x-y; m) \;.$$

We give a brief discussion of charge conjugation in this theory. The *charge conjugate field* $\phi_c(x)$ is given by (C.30) with $a_i(\vec{p})$ replaced by $b_i(\vec{p})$; $b_i{}^*(\vec{p})$ by $a_i{}^*(\vec{p})$; i.e., $\phi_c(x)$ is $\phi(x)$ with the roles of particles and antiparticles reversed. Now any element $\Phi$ in $\mathcal{H}^N$ may be written

$$\Phi = \sum_{r,s=0}^{\infty} \Phi^{(r,s)} \;,$$

$$\Phi^{(r,s)} = \sum_{i_1\cdots j_s=1}^{m} \Phi^{(r,s)}_{i_1\cdots i_r, j_1\cdots j_s}(\vec{p}_1\cdots\vec{p}_r, \vec{q}_1\cdots\vec{q}_s)\, u_{i_1}(\vec{p}_1)\otimes\cdots\otimes u_{j_s}(\vec{q}_s) \;.$$

We may define a charge conjugation operator $U_c$ by

$$[U_c\Phi]^{(r,s)}_{i_1\cdots j_s}(\vec{p}_1\cdots\vec{q}_s) = (-1)^{\sigma rs}\,\alpha^{r-s}\,\Phi^{(s,r)}_{j_1\cdots j_s\, i_1\cdots i_r}(\vec{q}_1\cdots\vec{q}_s, \vec{p}_1\cdots\vec{p}_r) \;,$$

where $|\alpha| = 1$; then

$$U_C a_i(\vec{p}) U_C^{-1} = \alpha b_i(\vec{p}) ,$$

$$U_C b_i(\vec{p}) U_C^{-1} = \bar{\alpha} a_i(\vec{p}) .$$

Thus we finally have

$$U_C \phi(x) U_C^{-1} = \alpha \phi_C(x) .$$

For our purposes the phase $\alpha$ is arbitrary. Notice that since (C.8) implies

$$u_i(\vec{p}) = C \bar{v}_i(\vec{p}) , \tag{C.33}$$

we have the additional relation

$$\phi_C(x) = C \phi^*(x) .$$

(3). Self-charge-conjugate theory.

The equation (C.1) may also be quantized to give a self-charge-conjugate theory, in which each particle is its own antiparticle. Thus the Hilbert space here is just the space $\mathcal{H}$ discussed in (1). The field $\phi(x)$ is defined by

$$\phi(x) = (2\pi)^{-2} \int d\Omega(\vec{p}) \sum_{i=1}^{m} \{a_i(\vec{p}) u_i(\vec{p}) e^{-ip\cdot x} + a_i^*(\vec{p}) v_i(\vec{p}) e^{ip\cdot x}\} .$$

Just as before, $\phi(x)$ satisfies the equations (C.31) and (C.32); however, $\phi$ satisfies the additional equation

$$\phi(x) = C \phi^*(x) , \tag{C.34}$$

obtained using (C.33). Thus $\phi(x)$ does not commute (anticommute) with itself; its commutation relations with itself may be derived from (C.32) and (C.34). Note that if we phrase our definition of the charge conjugate field as "$\phi_C(x)$ is $\phi(x)$ with the particle and antiparticle operators exchanged," it also applies to the theory considered here and we have

$$\phi_C(x) = \phi(x) ;$$

this is the origin of the term "self charge conjugate".

## BIBLIOGRAPHY


[1] H. Araki, *J. Math. Phys., 2, 267* (1961).

[2] N. N. Bogoliubov and D. V. Shirkov, *Introduction to the Theory of Quantized Fields* (Interscience Publishers, Inc., New York, 1959).

[3] N. N. Bogoliubov and O. S. Parasiuk, *Acta Math., 97,* 227 (1957).

[4] C. G. Bollini, J. J. Giambiagi, and A. Gonzalez Dominquez, *Nuovo Cimento, 31,* 550 (1964).

[5] E. R. Caianiello, *Nuovo Cimento, 10,* 1634 (1953).

[6] E. M. Corson, *Introduction to Tensors, Spinors, and Relativistic Wave-Equations* (Blackie and Son, Limited, London, 1953).

[7] F. J. Dyson, *Phys. Rev., 75,* 486 and 1736 (1949).

[8] R. J. Eden, P. V. Landshoff, D. I. Olive, and J. C. Polkinghorne, *The Analytic S-Matrix* (Cambridge University Press, Cambridge, 1966).

[9] H. Flanders, *Differential Forms with Application to the Physical Sciences* (Academic Press, New York, 1963).

[10] N. E. Fremborg, *Proc. Roy. Soc. (London) A188,* 18 (1946).

[11] I. M. Gel'fand, R. A. Minlos, and Z. Ya. Shapiro, *Representations of the rotation and Lorentz Groups and their Applications* (The MacMillan Company, New York, 1963).

[12] I. M. Gel'fand and G. E. Shilov, *Generalized Functions, Vol. I.* (Academic Press, Inc., New York, 1964).

[13] ———, *Veralgemeinerte Funktionen II* (VEB Deutscher Verlad der Wissenschaften, Berlin, 1962).

[14] M. Gell-Man and F. Low, *Phys. Rev., 84,* 350 (1951).

[15] T. Gustafson, *Arkiv. Mat. Astron. Fysik, 34A, No. 2* (1947).

[16] R. Haag, *Dan. Mat. Fys. Medd., 29,* 12 (1955).

[17] ———, *Phys. Rev., 112,* 669 (1958).

[18] K. Hepp, *Commun. Math. Phys., 2,* 301 (1966).

[19] P. J. Hilton and S. Wylie, *Homology Theory* (Cambridge Univ. Press, Cambridge, 1962).

[20] A. Jaffe, *Commun. Math. Phys. 1,* 127 (1965).

[21] G. Källen, *Arkiv Fysik, 5,* 130 (1951).

[22] E. Karlson, *Arkiv Fysik, 7,* 221 (1954).

[23] H. Lehmann, K. Symanzik, and W. Zimmerman, *Nuovo Cimento, 1,* 205 (1955).

[24] N. Nakanishi, *Graph Theory and Feynman Integrals* (to be published by W. A. Benjamin, Inc.).

[25] E. Nelson, *Operants: A Functional Calculus for Non-Commuting Operators* (to be published in the Proceedings of the Conference in Honor of Marshall Harvey Stone).

[26] S. B. Nilsson, *Arkiv Fysik 1,* 369 (1950).

[27] O. S. Parasiuk, *Ukr. Math. J., 12,* 287 (1960).

[28] M. Riesz, *Acta. Math., 81,* 1 (1949).

[29] A. Salam, *Phys. Rev., 82,* 217; *84,* 426 (1951).

[30] M. Shensa, *Renormalization by Distributions* (Princeton Thesis, 1966).

[31] L. Schwartz, *Théorie des distributions* (Hermann & Cie., Paris, 1966).

[32] S. Schweber, *An Introduction to Relativistic Quantum Field Theory* (Harper and Row, New York, 1961).

[33] E. Speer, *J. Math. Phys., 9,* 1404 (1968).

[34] R. F. Streater and A. S. Wightman, *PCT, Spin and Statistics, and All That* (W. A. Benjamin, Inc., New York, 1964).

[35] H. Umezawa, *Quantum Field Theory* (North-Holland Publishing Company, Amsterdam, 1956).

[36] Y. Takahushi and H. Umezawa, *Prog. Theor. Phys., 9,* 14 and 501 (1953).

[37] G. N. Watson, *A Treatise on the Theory of Bessel Functions,* Second Edition (Cambridge University Press, Cambridge, 1962).

[38] S. Weinberg, *Phys. Rev., 80,* 268 (1950).

[39] G. C. Wick, *Phys. Rev., 80,* 268 (1950).

[40] P. Breitenlohner and H. Mitter, *Nuclear Phys., B 7,* 443 (1968).